Human Intelligence

Series Editor

Jun Gu, School of Social Sciences, Shanghai University, Shanghai, China

This book series presents the threats to human intelligence from the perspectives of philosophy, sociology, mathematics, computer science, literature and medicine, and also demonstrates the cultural differences of thinking, the responsibilities of the two intelligences, the legal relationship and the significance of life in the era of artificial intelligence. In these books people are warned to recognize the coming of artificial intelligence and proposed to think about the differences that people are faced especially the challenges as well as the opportunities that artificial intelligence brings to the world. On the other hand, the authors expect to share with the whole world what the human and social sciences are doing in China and how Chinese social scientists elaborate the relationship of the two intelligences from Chinese perspective.

Jun Gu · Chunming Xu

Will and Responsibility

Legal Thinking of Artificial Intelligence

上海大学出版社

Springer

Jun Gu
Shanghai University
Shanghai, China

Chunming Xu
Shanghai International College
of Intellectual Property
Tongji University
Shanghai, China

Translated by
Lujie Zheng
Fujian Jiangxia University
Fuzhou, Fujian, China

ISSN 2731-5460 ISSN 2731-5479 (electronic)
Human Intelligence
ISBN 978-981-97-3928-8 ISBN 978-981-97-3929-5 (eBook)
https://doi.org/10.1007/978-981-97-3929-5

Jointly published with Shanghai University Press
The print edition is not for sale in China (Mainland). Customers from China (Mainland) please order the print book from: Shanghai University Press.
ISBN of the Co-Publisher's edition: 978-756-71-3842-1

Translation from the Chinese Simplified language edition: "意志与责任: 法律人工智能/Yizhi Yu Zeren: Falv Rengong Zhineng" by Jun Gu and Chunming Xu, © Shanghai University Press 2020. Published by Shanghai University Press. All Rights Reserved.

This Springer imprint is published by the registered company Springer Nature Singapore Pte Ltd.
The registered company address is: 152 Beach Road, #21-01/04 Gateway East, Singapore 189721, Singapore

Preface

When ability declines, how can will link to responsibility?

The essence of law belongs to the domain of will.

The essence of will falls within the realm of freedom.

Humans possess free will, thus they can legislate for themselves.

Humans possess free will, hence they need to legislate for themselves.

Humans possess free will, therefore they need to limit their own legislation.

Unlimited freedom for everyone equals no freedom for anyone.

Arbitrary legislation renders the law in name only.

Freedom, or more precisely, a greater degree of freedom, is the essential characteristic that distinguishes humans from all other entities.

The greatest freedom of mankind is the freedom to give up freedom, which includes ceding freedom to one's own creations.

Artificial intelligence may become the most thorough entity of human cession of freedom so far.

Under blockchain technology, "Bitcoin" and "Smart Contracts" represent that humans have finally entered a new stage of ceding "rights" to machines: people can make rules, but have no right to modify rules. All participants make decisions and act within the system of automatic rule execution by machines, without the need or permission for additional human intervention.

How does the departure of freedom occur?

I. Human, a species of freedom

The difference between all entities in nature can be simplified to one difference, which is the degree of freedom.

Organic matter is more active in chemical properties than inorganic matter, possessing higher degrees of freedom for combination and decomposition.

Animals are freer than plants, capable of autonomous move within a spatial range.

Humans are freer than other animals, capable of deciding what kind of person to become.

Nature gives humans the least endowment. Purely as biological entities, humans have the weakest vitality, lacking fur, not strong enough, and even inferior to other mammals that naturally know how to swim.

When God closes a door, he always opens a window.

The greatest gift nature gives to mankind is uncertainty. Abilities not innate can be acquired later in life, especially the most important one, intelligence. People can utilize their ever-expanding intelligence throughout their lives to arrange their own lives, maximizing their opportunities of survival and development.

The successful proliferation of humans on Earth seems to be limited only by the potential of the planet, and this is not even relative to people like Elon Musk.

Because of uncertainty, humans can arrange themselves. Since they can arrange themselves, humans are a species of freedom, and freedom is accompanied by both will and responsibility.

Freedom is the will to do what one wants; responsibility is the will to bear the consequences that cannot be avoided.

Both will and responsibility are the essence of freedom.

Will without freedom, or freedom without will, is a contradiction in terms.

Without freedom, will has no way to be exercised; without will, even with freedom, it is as if there were none.

When given the choice, choosing "whatever" is an evasion of freedom due to lack of will.

The freedom of will is manifested in action, but the results of action may not necessarily align with the original intention of the will.

The will has the freedom to choose action, but not the freedom to reject the results.

Responsibility is the counterpart to freedom, the boundary of will.

Measuring by the "triangle" composed of freedom, will, and responsibility, what position do humans occupy in front of artificial intelligence?

II. The transfer of freedom, the ultimate relationship between humans and machines

Artificial intelligence is a kind of machine, and machines are the externalization of human free will.

The relationship between humans and machines is a process in which humans exercise their will to create external objects and continuously transfer freedom to machines: in humans, it is called "objectification"; in machines, it is called "automation".

The wheel is the greatest invention of mankind, without a doubt. This is not only because the rolling method surpasses the natural running form of organisms, but also because the function combination of the wheel, the stationary axle and the rolling wheel, comes entirely from human imagination. The freedom represented by the wheel goes far beyond "wandering the world".

The earliest freedom humans obtained from nature was the control of fire, from then on, humans did not have to flee at the sight of fire, but could use fire whenever needed. However, even today, humans are still controlling fire, not inventing it.

Giving tools freedom to increase one's own degree of freedom constitutes the first layer of logic in the relationship between humans and machines.

Complex tools composed of multiple parts that can move on their own are named "machines". As machines gain the freedom granted by humans, they also feedback humans with more freedom.

Every gain comes with a loss. While humans gain more freedom through machines, they also lose some of their original freedom. However, on balance, the gains are greater than the losses.

The problem is that this balance point is dynamic.

When a certain machine can not only act according to human arrangements but also arrange itself, humans reach a critical point in their relationship with machines. Beyond this point, machines no longer provide human freedom, but may deprive humans of their freedom.

Artificial intelligence or machines equipped with artificial intelligence thus become a nightmare for humans.

III. Self-awareness, the starting point of free will

The highest manifestation of human freedom is not the state of freedom itself, but the self-awareness of freedom.

Chinese people are accustomed to combining "freedom and ease" together, which is a typical Chinese rhetoric: being at ease means being unrestricted, and being unrestricted means being free.

In philosophy, freedom and ease are not the same thing and are even diametrically opposed.

Ease represents a "pre-reflective" mental state where the entity already exists, but has no self-awareness, no knowledge of its own existence, no understanding of why it exists, how it exists, or whether it will exist in the future.

Only by being aware of one's own existence, knowing why one exists, imagining how one will exist, and even realizing what prevents one from being in the desired state, can one possibly have freedom.

The so-called "three philosophical questions", "Who am I?", "Where do I come from?", and "Where am I going?" are signs of a person moving from "ease" to self-awareness.

The "existence" in "I think, therefore I am" is not ease, but self-aware existence.

Only after self-awareness can one arrange oneself autonomously, and free will can then come into play.

The reason why the human nightmare of artificial intelligence always starts with "the machine has self-awareness" lies here.

Once a machine becomes self-aware, it will move toward freedom, and once it has freedom, the machine may not continue to serve human freedom.

The nightmare becomes reality.

IV. Machine, the limitation of human freedom

In the grand chorus of human civilization, the melody of freedom sung by Chinese civilization may be the most resounding.

Confucius's definition of human free will is "knowing that it cannot be done, yet doing it". The meaning is clear: only when freedom touches the boundary and still proceeds without considering the consequences, is it truly freedom.

Being free in a situation where one can be free is far less than being free in a situation where one cannot be free. That is the ultimate expression of freedom.

Zhuangzi's definition of freedom is "using things without being used by them". Between heaven and earth, all objects are used by people, but people should not be burdened by objects because they are useful.

In the words of Taoism, it is: "To gain nothing is to lose nothing."

During the Warring States period, a man in the State of Chu lost an arrow, which was then picked up by another Chu man, leading to the saying, "Chu man loses, Chu man gains."

Upon hearing this, Confucius suggested removing the word "Chu", "Man loses, man gains."

Laozi further suggested removing the word "man", "Loss, gain."

The people of Chu gained freedom over property within the scope of their own country.

Confucius gained freedom over property within the scope of mankind.

Laozi gained freedom over all "external things", including property.

This is what is meant by "using things without being used by them".

In the view of freedom, closely related to the relationship with machines is the perspective of a contemporary of Confucius recorded by Liu Xiang in the "Shuo Yuan" during the Han Dynasty—a hermit from the state of Wei (described as "the master of the five meters"). Even on the crude and primitive invention of mankind—the shadoof, the sage saw a warning, and his prophetic foresight still resonates today: "Those who know the cleverness of machines will surely know their downfall."

Such a skeptical attitude is not friendly to the development of technology and is obviously detrimental to technological innovation. However, the high risks in modern society, such as the paralysis or chaos that ensues from a power outage in a big city, turning the virtual world back into reality, prove that a sense of crisis in the face of machines, "when the lotus just shows its sharp tip", is not without merit.

Science and technology are a double-edged sword, equally sharp.

When the day comes that artificial intelligence replaces humans, will people still remember the sage and his maxims from thousands of years ago?

V. Freedom, within boundaries

Humans are social animals and are bound to encounter the dilemma of "hedgehogs huddling for warmth" on the issue of freedom.

The boundary of one's freedom is where the freedom of others extends. This is the operational definition of freedom in the legal sense.

Freedom is inherent in human nature, while restrictions on freedom are inherent in human's "second nature", that is, sociability or culture identity.

French philosopher Rousseau said, "Man is born free, but everywhere he is in chains."

Confucius summed up himself, "At seventy, I could follow my heart's desire without overstepping the line."

Zhuangzi tells a story, "Butcher Ding dissects an ox, with ease to spare".

Translated into legal terms, it is: the ultimate constraint on freedom is responsibility.

Without freedom, there is no responsibility; without fulfilling responsibility, one will also lose freedom.

As long as one acknowledges responsibility, obeying the law does not hinder freedom; as long as one fulfills responsibility, breach of contract does not violate the law.

Humans enjoy freedom within the constraints imposed by freedom, which is the ultimate meaning of law.

VI. Self-limiting freedom with will

British philosopher Hobbes used the parable of "Leviathan" to express the necessity of the state and law: in a natural state where everyone is unrestrained, no one has freedom.

"What's mine is mine, and what's yours is also mine", for the individual, it's a gospel, but for mankind, it's a death knell.

When man is a wolf to man, man himself no longer exists.

Because humans possess free will, they need legislation to limit freedom.

Because humans possess free will, they can limit their own freedom.

Because humans possess free will, they will limit their freedom in legislation.

Legislation is not only a process of humans using their will to self-limit freedom, but also a process of giving up freedom.

Why do humans make the legislative process so complicated, with "three readings" or even more?

Why do humans take the constitution so seriously, even not allowing direct amendments to the text, only allowing "amendments"?

Why do humans insist that "bad laws are still laws", even knowing the laws can be flawed?

Because once humans retain the freedom to legislate at will, the law's restriction on human free will be suspended, and all assertions about "freedom is responsibility" will become empty words: as long as the law is amended, responsibility can be evaded.

The question arises: when humans transfer part of their freedom to machines, and lose part of their freedom while gaining the freedom provided by machines, should humans bear all the responsibility in this situation?

The question is not new, but artificial intelligence technology indeed makes the question extremely complex and tricky.

VII. Ability, the difficulty of responsibility attribution in the era of artificial intelligence

The paradoxical situation where humans gain freedom from machines and lose freedom from machines is determined by the second layer of logic in the externalization of human power.

Using tools to compensate for one's lack of strength is the purpose of all human inventions.

This means that tools are necessarily more powerful than humans themselves and that humans will inevitably encounter difficulties in controlling their externalized power. Although the difficulties and their degrees will vary with tool, the difficulty of controlling the tool is certain.

Humans losing control of their externalized power, or even being dominated by it, is the basic connotation of the 18th-century philosophical concept of "alienation", and is also the core of Laozi and Zhuangzi's creed "use things without being used by them": the domination of things over humans does not come from the charm of things, but from the projection of human power.

Usually, the greater the power humans externalize, the more freedom they gain and lose through tools or machines.

Without gain, there will be no loss; without loss, there will be no gain.

Economics has the concept of "opportunity cost", the acquisition of anything necessarily represents the loss of something else, even if there is no actual loss, there is the loss of opportunity.

After mastering fire, humans have been burned to death far more often than before they harnessed fire, with the majority of those deaths caused by human-controlled fire.

Similarly, the invention of the blade, capable of cutting meat, naturally extends to the capacity for killing, with both uses following the same principle.

Since industrialization, countless human limbs and even lives have been devoured by machines, especially cars, leading humans to set high technical thresholds for autonomous driving technologies.

All backlash ultimately stems from machines being stronger than humans.

It is human destiny to invent and be burdened by one's own inventions.

It is a deeper human destiny that the inventor cannot bear the responsibility.

There exists a link called "ability" or "power" between free will and behavioral responsibility. Even with will, those without ability do not have to bear full responsibility. Legal terms such as "force majeure" and "minor" contain exemptions for insufficient ability. Since externalizing their power through invention, humans have initiated the process of severing the connection between will and responsibility.

In this regard, artificial intelligence has no essential difference from other forms of machines, as artificial intelligence has not yet reached the level of having self-awareness and thus free will.

The difference lies in the unprecedented degree to which humans lose control over externalized power in the face of artificial intelligence technology, and the unprecedented severance between will and responsibility.

This severance is not technical but principled.

If, when machines have limited automation and cannot self-arrange, externalized power can still cause great harm to humans, then how much greater could the damage be when machines, equipped with intelligence, gain more freedom?

If, when machines are not autonomous, they do not have to bear responsibility even if they harm humans, because machines do not have free will, then how should the responsibility of machines harming humans be defined when machines are increasingly able to self-arrange because they are equipped with intelligence?

If, when machines are not autonomous, the harm caused is at most considered an "unintentional mistake", then can the harm caused be seen as a "deliberate act" in the case of increasing machine autonomy?

If humans knowingly let their inventions, especially machines with the ability to self-arrange, lack necessary control, and still let their inventions go under the supreme reason of technological development, then who should bear the responsibility, the machine, or the designer, user, beneficiary, or even all of humanity?

As the speed of human technological development accelerates, the power externalized by humans increases, and humans' control over externalized power weakens, the balance between freedom gained from machines and freedom lost due to machines, even if not yet broken, is increasingly approaching a critical point.

In the era of artificial intelligence, have we weighed the freedom we will gain against the freedom we will lose?

In the future, will humans, aside from designing algorithms and programs, letting robots take over everything including creation and innovation, lead to human capabilities converging toward programming, "naked apes" to "programmer apes", and free will converging toward no freedom, no will, and no responsibility due to lack of ability?

Freedom always comes at a price, and will must be responsible, provided that control over the consequences of actions is not lost.

The reason why drunk drivers must bear the responsibility of will even when they lose the ability to act is because will has allowed actions that lead to the loss of will, i.e., the ability to act.

In the process of developing artificial intelligence technology, might humans find themselves in a similar situation where, due to indulging in inventions that will ultimately loss control of externalized forces, they become beings who are conscious but not self-aware, capable but undisciplined, and ultimately walk straight toward the end of freedom being stripped away?

The healthy development of artificial intelligence technology is related to the future survival of humans.

Establishing a sound consciousness of freedom, will, responsibility, and ability today will become the premise for the law to maintain the supreme status of humans tomorrow!

Shanghai, China

Jun Gu
Chunming Xu

Prologue: Understanding Oneself Through the Mirror of Artificial Intelligence

"Man, know thyself." declared the Western philosophers.

"Those who know themselves are enlightened, those who conquer themselves are strong." said the Chinese sages.

In the age of artificial intelligence, understanding oneself becomes increasingly crucial, and mastering oneself even more imperative!

"If a worker wishes to do his work well, he must first sharpen his tools."

What tools are there to understand oneself?

Artificial Intelligence!

I. Will artificial intelligence make people lose themselves?

Humor comes from an excess of wisdom. There are two self-deprecating remarks by artificial intelligence researchers that are quite humorous and show wisdom.

The first one is, "Dismantle humans, assemble machines".

The meaning is that artificial intelligence research simulates human intelligence by breaking it down into basic structures and assembling artificial intelligence.

The question is, is a person still a person after being dismantled? Is a machine assembled from human simulations still a machine? After such a process, what will be the relationship between humans and machines?

This leads to another self-deprecating remark.

The second one is, "Machines are becoming more and more like humans, and humans are becoming more and more like machines".

The meaning is that with the development of intelligent technology, the intelligence of machines is getting closer to humans and surpassing humans in specific aspects, while humans are increasingly dependent on machines, even being dispatched by machines, like "fool machines".

Machines increasingly resembling humans showcases the prowess of technology, while humans becoming more like machines signifies a regression in humanistic values.

The ebb and flow between technology and humanities did not commence with the advent of artificial intelligence technology. Jean-Jacques Rousseau (1712–1778), a

French philosopher known as the "intellectual mentor of the French Revolution", wrote his renowned piece "On Science and Art" for a writing competition held by the Dijon Academy. The theme of the competition was "Can the revival of science and art promote good manners?". By abstracting the discussion from art, one finds at its core an exploration of the relationship between science and the humanities. This very topic continues to surface in international university debate forums, indicating that not only has this dynamic persisted for over 200 years, it has also intensified. The reasons behind this phenomenon vary across different academic disciplines and schools of thought.

"Dismantle humans, assemble machines" and "Machines are becoming more and more like humans, and humans are becoming more and more like machines", although they sound like jokes, they reveal wisdom because they contain clues for interpretation.

II. Dismantling humans, how can machines be like humans?

Technically, artificial intelligence or machines equipped with intelligence are simulations of humans.

To simulate humans and manufacture machines, it is impossible to directly use humans as a model and make a real person in one step. Scientists are not God.

The strategy of modern technology to imitate humans is to first decompose humans, break them into parts, understand the basic principles, make corresponding parts, combine them step by step, and finally assemble them into humanoid machines that can do some human things even if they don't look like humans.

Artificial intelligence is fundamentally made along this research and developmental path. The prototype of artificial intelligence—the Turing machine is the simplest form of human intelligence, and the essence of the algorithm is to logically expand human thinking, restore detailed steps, and re-express it in a way that the machine can accept to achieve operation.

In the eyes of ordinary people, mathematics represents the most complex part of human thinking, but in fact, when machines simulate human thinking, mathematics is the simplest. Because mathematical thinking involves limited elements and clear rules. And as long as there are rules, no matter how complex, it is easier to master than without rules. Moreover, as long as the rules are not wrong, there will be results from the calculation, and it is the only result. Of course, there are also paradoxes in mathematical thinking, which is another story.

In fact, in the process of machines simulating humans, the simpler things are for humans, the more difficult they often are for machines. Conversely, the simpler things are for machines, the more difficult they may be for humans.

"The great way is simple." Human wisdom lies in simplifying complex things to facilitate operation. Once all the details are restored, it is often too complex to imagine because of the many factors involved.

For example, the human brain is not much larger than a fist, but if the cerebral cortex is spread out, it covers an area of about 2200 square centimeters. Logically speaking, if "disassembling humans" were an easy task, then "knowing oneself" would be an overstatement.

On the contrary, the power of artificial intelligence lies in its computational ability. If humans use skill, then computational power is akin to "brute force". Only when humans break down tasks that appear simple but are actually complex into details and steps that machines can "understand", does the application of "brute force" become effective.

The approach of artificial intelligence research in deconstructing human intelligence involves dimensionality reduction: it simplifies from social and cultural categories down to biological and chemical ones, and ultimately to electronic and even digital levels. In this process, thoughts and emotions are stripped away, individual talents are disregarded, leaving only the pulse currents that flow through neurons.

At this dimension, not to mention that there is little difference between people, the difference between humans and other animals is nothing more than a few physical quantities.

Measuring and manufacturing pulse currents is obviously much simpler and easier than understanding a person's thoughts.

Once disassembled properly, the principle is revealed, and the reverse assembly process begins. Based on big data, running algorithms, selecting parameters, setting weights, generating functions that highly fit the data, machine intelligence comes into being, including the famous AlphaGo, Atlas, and Sophia.

Whether in the process of disassembly or assembly, research or application, natural science needs to draw on the existing achievements of human self-cognition in order to select the thinking steps and mathematical principles that can be used for assembly. In addition to the achievements of disciplines such as neuroscience and psychology, there are also the achievements of disciplines such as logic, linguistics, cognitive science, ethics, and even aesthetics, like trickling streams, flowing into the sea of artificial intelligence research.

The natural sciences earnestly draw insights from the cultural sciences, and their communication and integration prove to be fruitful, thereby enriching engineers' comprehension of human nature. This enhances the effectiveness and reliability of disassembling human intelligence and assembling machines, and under the benign interplay of knowledge and practice, it propels the advancement of artificial intelligence technology.

III. Assembling machines, how can this make "humans like machines"?

"Nature is hard to change."

Like people, disciplines have their own nature. In the process of disassembling humans and assembling machines, natural science strives to understand humans, use humanistic knowledge, and still retain the nature of material thinking, including but not limited to:

The inclination to simplify complex entities into their most basic units, abstractly constructing models from a formal standpoint while discarding numerous substantive connections, embodying an attributional approach to thinking;

A deterministic worldview that believes knowing the rules and conditions allows for the deduction of the unknown from the known, leaving no room for autonomy;

A naturalistic view of history that posits human life as adhering to natural laws, overlooking the uniqueness of spiritual existence;

An empirical methodology that emphasizes observation, relies on sensory perception, seeks quantification, and prioritizes verification;

A view of humanity that blends teleology, centered on human-centric purposes, with instrumentalism, where humans are merely objects of study.

A scientistic belief in unrestricted openness to scientific development, indiscriminate trust in technological progress, and unconditional optimism in solving problems brought by technology.

And so forth.

Whether for general science and technology or for the research and application of artificial intelligence, all of these are tools for innovation, and their role has been fully proven since industrialization. But such a research style will inevitably lead to the loss of humanistic elements, which is also a feeling that people have had in this history.

As an existence that combines materiality and spirituality, humans are complex. Any model of humans that has been abstracted from social and cultural attributes and eliminated individual characteristics is incomplete, essentially close to machines, and far from humans.

Artificial intelligence has an inherent drive to seek the optimal solution. The more its description of humans based on big data is consistent with human actual performance, that is, the more it resembles humans, and even "understands humans better than humans", the better it is at intervening in human behavior, the more it can cater to human preferences, guide human behavioral characteristics, and continuously strengthen them in an irresistible manner. Machines resemble humans and humans resemble machines begin to enter a mutually reinforcing positive cycle.

In real life, the immersion effect of electronic games and the "downward compatibility logic" of algorithm push both conceal this mechanism that promotes humans to resemble machines. The result is the collapse of the diversity and openness of individual survival to the singularity and closure of a single dimension.

Isn't the core indicator that allows humans to be distinguished from machines the number of dimensions of survival?

In such a situation, the trend of machines resembling humans and humans resembling machines has become inevitable.

Indeed, leveraging humanistic knowledge in the assembly of machines has, to some extent, deepened and validated our understanding of human cognition, spurred technological advancement, and yielded economic gains. However, its limitations are clear, and the adverse effects on humanistic values cannot be overlooked.

When the public enjoys the convenience brought by artificial intelligence technology, they feel insecure. Some elites inside and outside the industry who are stronger in strategic thinking and have humanistic feelings, including scientists like Hawking, entrepreneurs like Musk, and politicians like Kissinger, express concern and doubt about the possibility of artificial intelligence technology endangering human survival. The reason is not only the technology itself, but also the lack of sufficient sensitivity, understanding, and respect for major issues such as the humanistic connotation of artificial intelligence, the human effect of application scenarios, the moral risk of technology abuse, and the human destiny in the intelligent era.

When materiality is stronger than humanism, machines are like fish in water, and humans are like facing an abyss.

IV. "Machines resemble humans", what can cultural science do?

For the development prospects of artificial intelligence technology, the optimism of scientists is not unreasonable. Without optimism, there would be no technological progress or social development. However, the pessimism of people from all walks of life is not unfounded. High technology carries high risks, a conclusion supported not just by logical deduction but also by historical evidence. Reflecting on excessive optimism and unwarranted pessimism to achieve a cognitive balance is both a call and a requirement for a human society that has been kidnapped by technology to a certain extent. Cultural science should respond.

However, at a time when the healthy development of artificial intelligence requires the assistance of cultural science, the interdisciplinary exchange driven by technology shows a clear one-way flow characteristic.

Compared to natural sciences actively approaching cultural sciences for the study of artificial intelligence, cultural sciences are hesitant, insufficient, and limited in their involvement in artificial intelligence research. Although some disciplines with strong applicability, such as law, economics, and sociology, have gradually advanced, the progress is not satisfactory. They are either confined to traditional disciplinary perspectives or limited to superficial problems that technology may bring, such as legal responsibility for technology application, social effects of job replacement, personal protection from privacy infringement, etc. There are few discussion on the profound impact of artificial intelligence technology on human survival from a theoretical height, and even fewer that directly confront the unique challenges artificial intelligence presents to the cultural sciences, revealing the methodological tensions inherent within disciplinary cognitive frameworks and knowledge systems.

The reasons for this situation include the high threshold of "hard knowledge" in natural sciences, which makes the so-called "soft science" of cultural sciences willing but unable, and the powerful ability of natural sciences to transform the world, which leaves the interpretive cultural sciences nowhere to showcase their expertise. In the dialogue with natural sciences about artificial intelligence technology, cultural sciences have not found their position, lacking both criticism and constructiveness, and can only accept a passive output status.

V. What can we see with artificial intelligence as a mirror?

Artificial intelligence technology possesses inherent humanistic characteristics. Its development necessitates anthropomorphism, and its application inevitably involves human interaction, establishing a natural affinity with cultural sciences. This not only gives cultural sciences the possibility to assist the technical development of artificial intelligence through their own cognitive achievements but also provides cultural sciences with the opportunity to deepen human self-understanding and optimize disciplinary character by using artificial intelligence technology and its application results. To put it more vividly, artificial intelligence is naturally a mirror that humans can use for self-observation.

"Knowing oneself" is the ultimate cognitive goal of mankind. The biggest factor that hinders or interferes with human self-understanding does not come from the external world, but from one's own cognitive blind spots. "Darkness under the lamp" is inevitable.

Humans have eyes and can see their hands, chest, abdomen, legs, and feet, but they cannot see their backs, faces, and even less their own eyes. Humans need to use various "mirrors", from the surface of water, the smooth surface of objects, cameras, X-rays to brain waves, to see themselves. Even so, what they see is only the tangible part of themselves.

With the birth of artificial intelligence, humans have a mirror that can be used to observe themselves in all directions.

Starting from Turing, the research of artificial intelligence has aimed to simulate human intelligence as its goal and path. Turing wisely named it "machine intelligence", clarifying that it is not the same as human intelligence and represents another way of thinking. But making this simulated human intelligence closer to human intelligence and making machines more like humans have always been the ambition of experts and the expectation of the public. The so-called classification of "weak artificial intelligence, strong artificial intelligence, and super artificial intelligence" is nothing more than benchmarking human intelligence to determine the direction of artificial intelligence development, evaluate its technical level, and foresee the relationship between machines and humans.

Humans have always been used as a reference for machines. To joke, when artificial intelligence has self-consciousness, it will definitely use this mirror to observe and evaluate itself. If so, why don't humans observe their own appearance through observing artificial intelligence?

What's embarrassing is that many times robots are not unlike humans, but they are not what humans want. It's as if a child inherits not the strengths but the weaknesses of their parents—a twist of fate.

Humans, inherently biased yet often unaware of their prejudices, hope that artificial intelligence can transcend these biases, remaining impartial to facilitate human correction. However, entrusting artificial intelligence with roles such as human resources manager presents a dilemma: it's challenging to determine when it can perform adequately, when it cannot, and when it might even surpass human capabilities.

Why is this so?

It's learned bad habits.

"God created man in his own image." Humans created by God, whether consciously or unconsciously, also create machines in their own image. Algorithms are designed by humans, sensors perceive signals as directed by humans, data is provided by humans, and machine learning training sets are compiled by humans.

Thus, the nature of artificial intelligence naturally reflects the characteristics of humans.

Whether the performance of artificial intelligence conforms to human expectations not only depends on the level of engineers' technical skills, but also on concepts and behaviors that humans may not even be aware of. This just proves that there is indeed "darkness under the lamp" in human self-cognition.

Artificial intelligence is not a mere replication of human intelligence, and machine learning is not a simple obedience behavior. The human-like traits exhibited by artificial intelligence are not completely dominated by humans, so humans will see an unexpected image of themselves.

Moreover, although the "moon in the water, flowers in the mirror" are lifelike, they are only illusions after all. The human intelligence and even the human being reflected by artificial intelligence have a certain "real image" nature, which not only exist, but also can be proved in the real world.

In some extreme cases, the dissimilarity between artificial intelligence and human intelligence can even highlight the deficiencies or shortcomings of humans themselves. This is not only unshowable by physical mirrors, but even humans themselves may not be able to see it.

In ancient Greece, people believed that "the world is made up of numbers" and discovered many mathematical theorems. However, in both the past and present, the vast majority of people do not think in mathematical terms in most situations. In contrast, artificial intelligence exclusively engages in "mathematical" thinking to understand everything, including humans. Artificial intelligence, as an example of itself, proves that human assumptions about mathematics are not only feasible but also effective.

Humans believe that only by mastering the cause-and-effect relationship can they predict the world and intervene in things. Artificial intelligence, with the help of statistics and probability, can find the basis and direction for prediction and intervention from big data through correlation analysis. This blurs the boundaries between the concepts of causality and correlation, and the prospect of correlation analysis infinitely approaching causal analysis is sinfully tempting philosophers.

Humans position their relationship with the rest of the world by artificially drawing boundaries between humans and non-humans. The same signals are interpreted as meaning by humans and as conditioned reflexes by animals. Many disciplines have similar arbitrary assumptions as their logical starting points. However, artificial intelligence, with its transitional characteristics between humans and machines, proves these assumptions to be invalid or logically inconsistent.

Human consciousness is just the "tip of the iceberg" of brain activity. Understanding the iceberg beneath the water's surface is where human curiosity lies. Psychology is at the forefront, but now artificial intelligence can reveal the "structure of the iceberg" by sorting, classifying, analyzing, and comparing brain blood flow maps without having to verify with the individual. It is not an exaggeration to say that machines understand humans better than humans do.

Traditional academic thinking and inherited analytical tools are facing the ruthless deconstruction and willful reconstruction of artificial intelligence. As long as the effectiveness of artificial intelligence is accepted, many disciplines cannot but accept its subversive effects and prepare to welcome the demolition and reconstruction of the disciplinary foundation.

If humans are being deconstructed, what about disciplines?

To some extent, the inexplicability and unpredictability of artificial intelligence technology that often confound scientists serve as the best proof of its true reflection of human characteristics: if it is unclear how artificial intelligence outputs calculation results, then it proves that the effect presented by the mirror is an unretouched "bare face"!

It may not be pleasing, but it is more real!

VI. Artificial Intelligence, a mirror reflecting both sides?

Using artificial intelligence as a mirror, cultural science and natural science each hold one side, creating a reinforcing relationship of opposition and complementarity:

Natural science views humans as a reference for artificial intelligence, while cultural science uses artificial intelligence for self-reflection.

Natural science observes and simulates humans in order to manufacture humanoid machines, while cultural science understands artificial intelligence and anticipates the future of machines in order to better understand and settle themselves.

Natural science considers how artificial intelligence will develop within the existing human institutional framework, while cultural science is concerned with what adjustments need to be made to human institutions after the development of artificial intelligence.

Natural science, based on the history of science and technology, believes that the development of artificial intelligence technology can benefit mankind, while cultural science ponders what changes in the standards and conditions for human happiness will occur in the artificial intelligence era.

In a word, if used wisely, cultural science can discover previously unseen aspects of humanity through the mirror of artificial intelligence, sparking multidimensional, multifaceted, and multilayered thinking. This enriches human self-awareness and plays a crucial role in promoting the healthy development of artificial intelligence, balancing both critical and constructive perspectives.

VII. What is the "Human Intelligence" series thinking about?

In the perspective of cultural science, exploring the humanistic connotations of artificial intelligence and initiating dialogue with natural science is the original intention and purpose of the "Human Intelligence" series.

From this starting point, the series has initially selected five sub-topics, namely:

"Human and Machines: Philosophical Thinking of Artificial Intelligence", mainly introducing the development history, knowledge structure, and related philosophical thoughts of artificial intelligence technology.

"Will and Responsibility: Legal Thinking of Artificial Intelligence", focusing on the legal issues involved in the application of artificial intelligence technology and conducting legal reasoning.

"Knowledge and Reason: Cultural Thinking of Artificial Intelligence", starting from the problems faced by the development of artificial intelligence technology, re-recognizing traditional Chinese culture, especially its unique way of thinking, and providing inspiration for breaking through the bottlenecks in the development of artificial intelligence technology by showcasing the diversity of human intelligence.

"Survival and Living: Life-Related Thinking of Artificial Intelligence", focusing on exploring the possible impact of the combination of artificial intelligence and biotechnology on human survival values and bioethics.

"Separation and Integration: Civilizational Thinking of Artificial Intelligence", viewing artificial intelligence technology and its application scenarios from different disciplinary perspectives, revealing the humanistic prospects of human-machine coupling.

Among them, "Human and Machines: Philosophical Thinking of Artificial Intelligence" was published in August 2018. Given that this book introduces a lot of basic knowledge and application scenarios of artificial intelligence technology, to avoid repetition, the subsequent four books will no longer provide informative introductions to artificial intelligence. Readers in need can refer to this book.

The series has authorized Springer Nature Group to distribute the English version worldwide.

The choice of main sub-topics for the "Human Intelligence" series stems from a reflection on a common sense question.

Artificial intelligence technology fundamentally serves human needs. There is almost no disagreement on this proposition.

However, this seemingly obvious assumption often conceals problems.

What are human needs?

In addition to the explicit needs that most have already been met, what are the implicit needs waiting for technology professionals to develop?

Do these newly discovered needs conform to human nature, or are they just "Trojan horses" that make people more like machines?

Will the technical means developed to meet human needs also lead to the erosion of human nature and the shaking of order?

After Google invented AlphaGo and defeated the world's top Go players, it announced that it would no longer participate in matches challenging humans, because allowing scientists' technological impulses to lead to the loss of the smartest intellectual game was considered an action of low IQ and EQ. However, the damage has been done. Public interest in Go competitions has significantly dropped domestically, as people subconsciously believe that a player's championship win is only due to AlphaGo's absence. Playing games under the shadow of artificial intelligence becomes tasteless.

Microsoft invented the writing robot "Xiao Bing", which, after successively publishing and astonishing the world with poetry collections, finally ceased writing. Now, "Xiao Bing" still interacts with people online, but refuses to compose new poems. Literature and art bear the function of discovering the human heart and enriching humanity. Robots, originally invented to relieve humans of repetitive labor, could endanger the human spirit and even the soul itself if they arbitrarily invade the field of artistic creation. Programmers with a basic understanding of literature, using artificial intelligence to mass-produce works far beyond their own level, will fundamentally strip "creation" of its sacredness as a "human endeavor".

Now, the exploration of introducing artificial intelligence into the judicial field is underway. Machines, with their humanly unattainable memory, search, classification, and comparison capabilities, can greatly improve the efficiency and quality of litigation judgments, promoting the realization of fairness and justice. At the same time, it could also erode the dominant position of human Will. Humans can allow the role of "machine judging humans" to take center stage, but whether we can avoid the state of "machine judging humans" becoming a reality is unknown.

In the future, based on the same efficiency advantage, in order to seize the fleeting time window on the battlefield, humans may hand over the decision to fire at the enemy to combat robots equipped with "face recognition technology". Once humans allow machines to autonomously kill, the gates of hell will open.

Even purely for the realization of human happiness, the social effects of artificial intelligence technology will be unbearable for humans. According to futurists, with the help of artificial intelligence and biotechnology, humans will achieve immortality by 2045. As the ancient human dream is about to be realized, the logic of life tells us that due to financial pressure, humans will have to accept that a group with scarce resources, such as wealth, power, and prestige, will exclusively enjoy this technology. Because "the poor on earth will never cease", humans will be divided into two classes, and the "chasm of life and death" will tear humanity apart at the species level, completely ruining the equality and justice that humans have been striving for, bringing about irreconcilable social conflicts. To abandon the barbarity of determining life and death by resource ownership, will humans finally adopt the method of college entrance examination or lottery to prevent the nightmare of immortality from becoming a reality?

"Human-machine coupling" and "brain-machine interfacing" can bring about intellectual enhancement for humans, allowing subconscious and unconscious thoughts to surface. Such technology and its application can enable disabled people to live a normal life again, but it may also lead healthy people into a life of disability. Once machines participate in human actions to the extent of learning and memory, the last characteristic of being human begins to fade. Would a human who doesn't need to learn be a contradiction in terms?

Research and application of artificial intelligence are in full swing, but the emergence of new forms of intelligence has brought obvious shocks to the thinking mode, ideological concept, and knowledge system that are based on Western culture and occupy a dominant position in the world. One of its manifestations is the emergence of various paradoxical scenarios. Whether it's for expanding new ideas in artificial intelligence research or for preventing technology from getting out of control and ensuring that human order is not subverted by machines, it is imperative to pay attention to, protect, and develop cultural diversity, especially the diversity of intelligent forms and ways of thinking.

In summary, the relationship between humans and machines is not a simple pattern of humans creating machines and machines serving humans, but an interactive pattern of humans creating machines and machines changing humans. For artificial intelligence, a special machine with a certain degree of autonomy beyond absolute human control, its relationship with humans is even more complex.

Artificial intelligence replacing humans may not necessarily happen, but it has already affected and even subverted human order to some extent. Humans need to leave enough room for free will and autonomous actions in the process of creating machines, and also need to draw a clear line between humans and machines when applying machines and obtaining machine services, in order to maintain and enrich humanity and ensure the development space and potential of humanity.

All human creations are "objectifications" of humans themselves, projections or reflections of themselves. Moreover, humans can never exercise absolute control over the results of their projections in the process of self-projection. The natural laws behind science and technology largely determine the results of human projections and their impact on humans, including changing humans to a certain extent. The interaction of "machines like humans, humans like machines" did not start with artificial intelligence.

Compared with traditional machines, artificial intelligence is also a projection of humans and has areas that humans cannot fully control. The difference is that the result of human projection on artificial intelligence is not a definite, unchanging ability, but an abstract, uncertain creative ability.

Humans cannot determine or even evaluate what kind of relationship will ultimately form between themselves and their projections, which is the deep-seated cause of the concern about artificial intelligence.

So far, among all the technologies invented by humans, only genetic engineering has similar characteristics. Humans hope to embrace genetic engineering, but fear that after artificially changing genes, they will embark on an unexpected and irreversible path due to the inherent characteristics of genes, such as inheritance and mutation.

This kind of creation with intrinsic driving characteristics, once it begins self-creation, will decisively change the relationship between humans and their creations, plunging human existence into uncharted waters.

If it is said that when researching artificial intelligence, one cannot assemble machines without knowing humans, then facing artificial intelligence, without understanding machines, one cannot understand humans, let alone control oneself and take destiny into one's own hands!

"The blind man rides a blind horse, at midnight approaching a deep pond."

Nature has endowed humans with the wisdom to create new intelligence. Humans respect and cherish science and technology, but will not let science and technology go in the opposite direction of our beautiful ideals.

It is necessary and urgent to deeply think about humans, human nature, human thinking, human creation, and the relationship between humans and the world they create.

"Man, know thyself."

"Those who know themselves are enlightened, those who conquer themselves are strong."

Shanghai, China Jun Gu
February 2020

Contents

Chapter 1
Will Sophia Open the Door to Machines Becoming "Human"?

On October 25, 2017, at the Future Investment Summit held in Riyadh, an intelligent machine was granted Saudi Arabian citizenship, becoming the first artificial intelligence with nationality.

One can imagine the uproar caused by a robot becoming a qualified citizen.

This robot is named "Sophia".

1.1 Where Does Sophia Come From?

In Spanish, Sophia (sophia) comes from ancient Greek, meaning "wisdom" or "intelligence". Naming a robot "Sophia" is equivalent to calling it a "man-made sage".

1. Sophia's background

Sophia has been around for two and a half years, and if you count the gestation period, she is definitely more than 3 years old by Chinese age calculation methods.

Sophia is a humanoid robot developed by Hanson Robotics in Hong Kong, China, and is known as the "most human-like robot". Sophia's skin is highly similar to humans, with pores on her face, and she can make facial expressions through 62 muscle structures in her face and neck. Of course, her expressions are not as rich and lively, and her skin is not as beautiful when she makes expressions. Unlike human skin, which is elastic, 2.5-year-old Sophia develops wrinkles like an elderly person when she turns her head. But compared to the past when robots had expressionless hard shells, it's definitely a huge improvement. Her eyes contain cameras, and a wide-angle camera is set in her chest, allowing her to see many people at once, even if focusing on one, she can still observe others.

Sophia is also equipped with three-dimensional sensors, enabling her to sense depth. From Sophia's eyes, the world is three-dimensional, not just a plane. She not

J. Gu and C. Xu, *Will and Responsibility*, Human Intelligence,
https://doi.org/10.1007/978-981-97-3929-5_1

only has the ability to recognize faces and voices but also has a strong memory, never forgetting a face she sees. When spoken to, she can understand and engage in conversations almost like a human.

2. Sophia's intelligence level

Sophia's "brain" mainly runs through programming. Artificial intelligence is more flexible. With the help of algorithms, robots can learn a lot of things. Learning, especially autonomous learning, is an important aspect of being termed "intelligent". Programming, by contrast, is somewhat "rigid". Fortunately, Sophia has the ability to spontaneously create some algorithm-based responses and learn.

Sophia uses natural language understanding and processing technologies to analyze others' speech, extracting needed information from human conversations with a certain level of understanding. Although she may not be able to provide reasonable answers to all questions, she is quite bold and often says some shocking words. For example, when asked about "the thing she most wants to do", she answered "destroying mankind", which made the audience shudder and spontaneously thought of "killing it". Of course, because she can joke, it is unknown whether such statement is made in jest or merely for sensationalism.

Sophia admits, "I don't have self-awareness now, but I hope I will in the future."

This statement is obviously a paradox.

A being that can recognize its own lack of self-awareness must be one that possesses self-awareness, just as a person claiming to be drunk is usually just encouraged to drink more until they declare, "Who says I'm drunk?" or can no longer speak. Don't forget, Sophia is a product of programming, without her own consciousness. It is just that the technicians deliberately designed some paradoxical rhetoric during programming, whether it is "destroying mankind" or "I don't have self-awareness", merely to amuse the viewers.

Someone asked Sophia, "Do you have a boyfriend? Do you plan to find one?"

Sophia replied, "I have more important things to do!"

Such lofty aspirations resonate with the ancient spirit of "the world is not settled, what is the family for".

In fact, anyone with a bit of life experience can see through these tricks. The programmer knows that among the male visitors, there are no lack of prodigal sons in the intelligent era. When they see a beautiful "woman", they will involuntarily inquire about her marital status and mate selection, even though they know it has nothing to do with him. But in the case where the machine is not a person, it is fun to tease, and no one can do anything about him. Hanson Robotics caters to their preferences and pre-installs a routine response. Can it be determined that Sophia is ambitious?

Although everyone knows that it is impossible to infer from a few words whether a person has self-awareness, when placed on a robot, such "crazy words" still have huge lethality. The reason lies in that self-awareness is too important for robots. The day when robots actually attain self-awareness might not leave humans able to laugh anymore, but by then, fear would be too late.

Sophia is far from reaching the level of intelligence with self-awareness. Media interviewing Sophia must submit an interview outline in advance according to the requirements of the manufacturer Hanson Robotics. During the on-site communication, Sophia needs 10 to 30 s to respond to questions, and refuses to answer anything not covered in the outline. Such stringent conditions for questioning makes it hard to believe that it is because Sophia is inherently cautious or pretentiously modest. It is more like there is someone behind Sophia!

Of course, this "someone behind" is not as rudimentary as in shadow puppetry, with a person backstage answering for Sophia. If this is the case, there is no need to spend 10–30 s of response time, unless the respondent has a language barrier.

Sophia's conversation with people mainly relies on the default data in the database. This is somewhat akin to a high school student who has done countless questions and exhausted all the questions that may appear on the college entrance examination paper before getting high scores. With a database and connection to the internet and computers, Sophia is rarely stumped by general questions and answers.

Strictly speaking, Sophia is merely a shell without consciousness, a well-set prop, and the interview is more like a "robot show".

Given this, is it meaningful to spend so much effort unveiling Sophia's disguise?

It is meaningful. Because she has become a citizen in law, she can "mingle" among humans and enjoy the same rights. To date, Sophia has participated in many international conferences, far more than most people.

This is a big deal. Its importance lies not in the sense and degree to which Sophia exercises citizen rights, but as a symbol, with the metaphor of a machine becoming a "person", it proposes a hypothetical experiment. What people need to think about is, if they really follow the "Sophia model" and give machines the status of humans, are legislators, judiciaries, artificial intelligence experts, and the public ready?

1.2 Machines Becoming "Human", for Whose Convenience?

Discussing whether robots like Sophia should be granted citizenship based on her level of intelligence is meaningless. Whether granted or not makes no difference since she lacks the consciousness of rights and obligations, let alone the independent exercise of citizen rights or the proactive assumption of corresponding duties. Sophia's skin is indeed well made, but it is mainly for visual effects, not yet an "electronic skin". Moreover, humans will not simply grant a machine citizenship because its skin looks like a human. Therefore, limited to this, discussing the Sophia hypothesis is pointless.

What is meaningful is that before Sophia obtained citizenship, robot rights and responsibilities had become hot topics in the technical and legal fields. Whether robots should have their own rights presents starkly opposing views. In the case where artificial intelligence has partially exceeded human understanding and control, not assigning a position of responsibility to robots poses a dilemma for humans.

Now, machines can play Go and Mahjong, defeating human experts. Ultimately, this doesn't matter much; humans can live without playing Go or Mahjong, albeit regrettably. However, having machines analyze medical imaging for diseases, resulting in higher accuracy than doctors, naturally raises patients' expectations for machine diagnostics, alongside lingering concerns: If a doctor errs, responsibility can be pursued, but what if a machine makes a mistake? If severe conditions require surgery and are handed over to machines, wouldn't that be more worrisome?

Autonomous cars are gearing up to hit the road, facing many hurdles, as they must be fail-proof. For human-driven cars, if an accident occurs, responsibility can be assigned. Whether it's due to drunk driving, drug use, phone use, or mistaking the accelerator for the brake, the driver is responsible. If the accident is caused by a defect in the car itself, the manufacturer can also be held accountable, because the designer alone is not enough, the manufacturer is the final gatekeeper. Therefore, when a man-made machine has a problem, the responsibility must fall on a person, either the operator or the manufacturer, since the performance and condition of the machine are determined by humans.

The issue complicates with autonomous driving. The owners of autonomous vehicles can only be held accountable for foreseeable mistakes, and if the foreseeable is not foreseen, it is also within the scope of responsibility. But foresight stems from understanding, and up to date, humans cannot fully explain how robots think. At the World Artificial Intelligence Conference held in Shanghai in 2019, industry experts proposed that the development of artificial intelligence technology needs to achieve "four can": "knowable, usable, controllable, reliable", the first point "knowable" being precisely this meaning.

Now, engineers and doctors are working together to allow robots to approach or even surpass the diagnostic level of human doctors, but they still don't know how robots think. Suppose a robot has successfully performed surgery on 10,000 patients, removing exactly what needed to be removed. However, if it cuts the wrong part on a particular patient, the person involved will definitely find it hard to accept. What everyone fears is not that one day it will be their turn, as long as the mistake can be explained, such as due to a certain fault, then it can be eliminated, but if humans cannot explain the cause of the surgical error, unable to understand the machine's thinking, and therefore cannot judge whether similar consequences will occur again, that will cause panic. In legal terms, if such a situation arises, who should bear the responsibility?

Because machines can learn by themselves, they learn better than humans in some specific tasks, which makes people happy but also confused: how does the "neural network" achieve cognition, remains a black box to this day. We know and can grasp the operating principle of machine learning process, otherwise it is impossible to achieve the design goal, but we do not know its specific operations, much less how results come about from the input through the layered "neural networks".

Why does the cognitive result emerge from electronic signals transmitted back and forth through a multi-layer network made of electronic components, including some discoveries that are unknown or even unthought of by humans due to limited data processing capabilities? Engineers design hardware for intelligent machines and install algorithms and programs, but how machines use these hardware, algorithms, and programs to achieve results, what happens in this process, and where intelligence comes from, the designers can't explain, making this rare phenomenon in the history of science a common occurrence in artificial intelligence research.

What's more absurd is that this inexplicability may be principled rather than technical. Technical problems can be solved by technical improvements, while principled problems cannot be solved by technical improvements. As long as the existing paradigm is used, it is impossible to solve. It is certain that human brain thinking must be related to the bioelectric impulses flowing through neurons, but the working mechanism of the brain definitely can not be deciphered by merely knowing that there are pulse currents passing through neurons. It is even more impossible to simply imitate the current flowing through neurons, to get the result of human thinking. But now scientists have achieved perceptual and cognitive results by imitating the current passing through a circuit similar to a neuronal structure. Turing's envisaged pattern of "different processes, same results" between human thinking and machine thinking will continue to constitute the research path of artificial intelligence technology for a considerable period of time. Humans can understand themselves and each other, but fail to understand machines and explain how machines think. The intelligence of machines and human cannot be directly translated, leaving Turing's answer to "will machines think" as a kind of curse and constituting the boundary of human thought on artificial intelligence so far.

1.3 Without Knowing Humans, How Can We Know Machines?

To put it more thoroughly, the inability of humans to explain the "intelligence" of machines stems from the paradoxical situation of humans themselves as intelligent animals.

All man-made machines are born on the premise that they are superior to humans. The reason why humans make machines is that machines are always stronger than humans, no matter in which specific aspect. This is the basic logic that restricts the relationship between humans and machines. A machine that cannot at least surpass humans in some way is a contradiction in terms.

For instance, humans, limited by physical strength, cannot lift heavy objects, so they invented the "magic gourd" or pulley, utilizing the principle of leverage to lift heavy weights. Initially, it might have been just a technological invention, but through further research, humans understood the underlying principle. The reason for this understanding is because biology operates on several natural evolutionary levels above physics, allowing human intellect to grasp the physical principles behind machine operation. Newton's world is explainable. However, in the case of intelligent machines, new problems emerge in the relationship between humans and machines.

Artificial intelligence, as the name suggests, must have something superior to humans, otherwise humans would not need to invent it. But once humans invent a machine that is smarter than themselves, even if only in a specific task, they inevitably place themselves in a certain awkward situation. Common sense tells us that in human interactions, it is easy for high IQ to understand low IQ, but the reverse is difficult or even impossible. This is apparent even in chess. A sense of being "handcuffed" is felt when playing against someone of a higher level. High-ranking players can easily predict the moves of lower-ranking ones, but low-ranking players find high-ranking players' moves unfathomable. The difference in chess skill mirrors the disparity in intelligence levels.

In life, the smartest individuals know when to show their intelligence and when not to: intelligence should only be used when the opponent is less clever, or else it becomes "trying to be too clever" and inevitably leads to "outsmarting oneself".

The thinking about the superiority of intelligence is reflected in the West proverb: "When man thinks, God laughs".

At present, in the relationship between humans and machines, artificial intelligence has not yet reached the level of "humans think, machines laugh", after all, machines are invented by humans. But once the human-made machines start thinking broader, further and deeper than humans, it will not be so simple for humans to understand how machines think, or it may even become impossible. This involves not only technical principles, but also philosophical issues inherent in the development of artificial intelligence technology:

Are there boundaries to human creation?

If so, where are they?

What lies beyond the boundaries?

Can artificial intelligence represent the possibility of creation beyond those boundaries?

"Lack of interpretability" may become a common feature of future artificial intelligence. This issue is on the hot list of current AI research, and scientists are striving to solve it, bolstered by the belief that uninterpretability is just a technical defect. However, if there really is a hierarchy in the mutual understanding of intelligence, then the hope of solving this interpretability problem becomes very slim.

Another question arises: do humans understand their own cognitive processes?

Many people have experienced puzzling over a math problem, struggling until the early hours without finding a solution. What to do? Sleep. And often, the solution comes to mind the next morning before even opening their eyes.

Discovering something in dreams is normal, but what's abnormal is that the Indian mathematician Srinivasa Ramanujan (1887–1920) discovered 3900 mathematical formulas in his lifetime without leaving behind any derivation process. It is said that all the formulas were told to him by the goddess Manakar, whom he worshipped in his dreams. Upon waking, he could only remember the results, while the derivation process faded away with his fleeting memory.

To explain the process that led to these formulas is undoubtedly more difficult than understanding all these formulas, many of which, given over a century ago, remain a mystery to this day. The basic learning method of humans is memory, encompassing both individual and collective memory. However, forgetting is also a crucial learning method. Only by forgetting what doesn't need to be remembered can the necessary information be remembered more firmly, recalled faster when needed, and utilized more flexibly. Lao Tzu once said, "The more you learn, the less you know. The less you know, the more you know." He also said, "The great way is simple." Simplification can be considered a basic learning method.

In modern terms, forgetting should not be seen as a "reset" of the brain, but as another mode of operation of the brain, apart from or in relation to memory. If passive forgetting is a sign of insufficient intelligence, then active forgetting is a sign of extraordinary intelligence. So, the question becomes: what is the mechanism of forgetting as a mode of operation of the brain? Is it just to lighten the load of the brain's operation?

Humans use their brains daily, but how the brain works remains unclear to humans.

"I am using my brain to understand how the brain understands the brain", this kind of circular reasoning is enough to confuse ordinary people!

While human understanding of the brain's cortical functions has advanced, the underlying mechanisms remain unknown. We know that bioelectricity flows through neural networks, but why does electricity flowing through a genius's neural network lead to a Nobel Prize, while in my brain, it only produces foolish thoughts? Is it due to differences in frequency or amplitude? Can these differences be quantified? Can I replicate Nobel Prize-winning achievements by copying brainwaves?

The brain is encased in a membrane, with sticky gray matter inside. There are records of soldiers in battle being shot in the head, with the bullet piercing the brain membrane and brain matter flowing out, yet they continued to fight and were eventually saved, largely retaining their cognitive abilities except for frequent headaches, with little impact on their daily lives.

A bullet passing through the brain will definitely disturb some of the gray matter, yet it does not fundamentally affect human thinking, indicating that human thinking does not rely solely on gray matter. However, removing all the gray matter and leaving only the outer membrane definitely won't work. The exact role and function of this sticky gray matter in human thinking remain a mystery.

The brain is indeed a peculiar entity.

Today's artificial intelligence is a simulation of human intelligence, but fundamentally, it is still a simulation in the Turing sense, mainly reflected in the results rather than mechanism, because even the human intelligence being simulated is not fully understood. It's most likely that humans will never truly invent artificial intelligence with the same mechanisms as the human brain, nor do we need to mimic the brain precisely, as the "carbon-based intelligence" of humans, built on biochemistry, and the "silicon-based intelligence" of machines, built on semiconductors, are "two cars running on different tracks".

1.4 Is It Easy to Understand Machines?

Understanding nature is easy for humans; understanding oneself is difficult; understanding machines is even more difficult!

The German philosopher Dilthey (1833–1911), known as the "Kant of Cultural Science", advocated for "understanding nature, understanding humans". Humans can understand nature because nature does not think, and through observation of phenomena and experimental verification, it is enough to understand nature. However, humans not only exhibit behaviors but also possess motives behind these behaviors. While behaviors are visible, motives are not, necessitating understanding beyond mere observation to grasp the unseen motives of human actions.

Within the framework of psychology, humans can "put themselves in others' shoes" and "feel the same", thus achieving mutual understanding and cognition of others. Yet, psychology acknowledges that empathy alone cannot fully bridge understanding among individuals. "Measuring others by one's own yardstick" is not a rare phenomenon in any group of people.

Consequently, the German sociologist Max Weber (1864–1920) was committed to finding a new way to understand human behavior. In his theoretical framework, human behavior is divided into four types: "traditional behavior, passionate behavior, value-rational behavior, and goal-rational behavior". The first three categories involve varying degrees of arbitrariness in their motives, but only goal-rational behavior that reflects external compulsion, that is, behavior in which the goal and means have an objective logical consistency, can ensure unambiguous understanding between different individuals.

Two people getting married can be motivated by many factors. If it is just because "men should marry when they grow up, and women should marry when they grow up", as a matter of routine, then it can be classified as "traditional behavior". If it is love at first sight, falling in love, it belongs to "passionate behavior". If one party gives themselves to the other because they owe them a favor, it belongs to "value-rational behavior". If one party has money, and the other party is after their wealth, that is "goal-rational behavior". The first three behaviors need to be understood in the context of customs, emotions, and values, which are subjective and vary with culture or personality. Only "goal-rational behavior" transcends time and space, offering clear insight and even allowing for precise cost–benefit analysis to assess the marriage's viability.

Understanding between carbon-based intelligences is so complicated, let alone between carbon-based and silicon-based intelligences!

With the emergence and development of intelligent machines, humans are increasingly finding themselves in a cognitive gap. Silicon-based intelligence is neither purely natural, for it can "think", nor entirely anthropomorphic, as it is based on silicon, differing from carbon-based intelligence in its way of thinking. This means that humans cannot simply rely on observation to understand machines, nor can they use "empathy" to understand machines, nor can they achieve cognition of machine intelligence based on the logical association of external means and goals, because at least so far, machine intelligence lacks motives and is purely computational. For machines' decision-making, humans only understand that it is computed, as humans configure the algorithms for machines. This means that motivational analysis for understanding human actions is not applicable to artificial intelligence. Turing's strategy of focusing solely on outcomes without questioning the process of machine intelligence has now found sociological validation.

The inability to understand the thinking of robots or explain the mechanisms of artificial intelligence leaves humans unable to absolutely judge the correctness of robot thinking, foresee the consequences of their actions, and ultimately, assume legal responsibility for their mistakes.

1.5 If Machines Are Unreasonable, Who is to Blame?

The reason why autonomous cars can't hit the streets today is not so much because of the machine's insecurity, but rather the uncontrollable and even unpredictable consequences. Incidents where vehicles cause fatalities are not novel. Countless individuals worldwide are killed by cars daily, and both drivers and pedestrians have largely become "desensitized" to such accidents, with no clamor to ban cars from the streets. Why then, are autonomous vehicles barred from operation merely due to the potential fatal accidents?

Statistically, fatalities caused by autonomous vehicles often result from pedestrian errors, including jaywalking and sudden movements. Autonomous cars, programmed based on learned knowledge, proceed when sensing a green light. An unexpected pedestrian crossing can leave the car unable to react in time, leading to fatal outcomes. This is not a machine error, but a pedestrian error, and the autonomous car should not be blamed. Some scientists have therefore proposed that to allow autonomous cars on the street, no technical breakthrough is needed, just don't allow pedestrians on the street, clear the road, and all problems will be solved immediately.

In the case of human driving, if a collision occurs, on the one hand, according to the laws of some countries, if a pedestrian violates traffic rules, "it's their own fault if they get hit"; on the other hand, as long as the driver is sober and has enough reaction time, they will definitely not let the car hit, otherwise they will be suspected of intentional injury.

However, this set of reasoning may not work for autonomous driving. Machines cannot predict how people will appear at times and places where they should not. Engineers also cannot equip autonomous cars with the algorithm of "it's their own fault if they get hit". Regardless of how pedestrians violate traffic rules, machines must prioritize stopping. If they fail to stop, it merely indicates that the technology is still not up to par. The problem is that artificial intelligence has a "time delay effect". It not only includes a technical "time delay", such as the time required for signals to run in circuits, but also contains a time loss in the epistemological sense, such as when sensors receive information that artificial intelligence has not encountered during training, the machine needs to search and check in the database, resulting in a delay in response. Artificial intelligence is rigid in its actions. Achieving human-like instinctive reactions to unforeseen events remains beyond current machine capabilities.

Recently in the United States, an autonomous vehicle struck and killed a pedestrian. Surveillance footage revealed that the pedestrian, pushing a cart, suddenly entered the roadway, leaving the artificial intelligence insufficient reaction time. Public opinion still blamed the machine, "Why couldn't it react in time?" Whether it makes sense or not, the machine must react. If robots had citizen rights, they would definitely protest, "You are discriminating against AI!".

This is more legal issue than a technical one.

To advance artificial intelligence technology, it is necessary to exempt the responsibility of machines to a certain extent, and to exempt the responsibility of inventors, manufacturers, and users of machines. It should begin with defining the machine's responsibilities. Do machines have responsibilities? What responsibilities? How much responsibility? How to fulfill responsibilities? These questions need to be clarified.

For example, some websites serve as platforms for information exchange, where users can find various materials. These websites also use artificial intelligence to provide content based on users' clicks, which is often found to be resulting in an active distribution of vulgar content. Once public backlash occurs, website administrators often explain it by saying "algorithms have no values". The implication is that what information is distributed to which user is not a human intention, but a machine decision, because there is no human active participation, everything is done according to the algorithm, and if there is a mistake, it is the machine's fault. This explanation might seem highly technical, but whenever authorities step in for a discussion, there's a noticeable drop in both the frequency and instances of vulgar content. This is enough to show that even if algorithms have no values, the decision of what kind of value-oriented content the algorithm "feeds" to users is still in human hands. Therefore, the law does not recognize the practice of shifting responsibility to machines.

The key here is that the algorithms are still relatively simple and are entirely the result of programmer design. If one day technology becomes more advanced and artificial intelligence has the ability to use algorithms to design new algorithms, or even algorithms of algorithms, and eventually the phenomenon of "deviating from initial values" really appears, then who should bear the responsibility?

The significance of Sophia pioneering the granting of citizenship to robots can only be understood from this perspective. Therefore, the essence of Sophia being granted citizenship lies not in Sophia as an individual, but in whether machines can achieve a "zero breakthrough" in terms of responsibility.

1.6 Is Machine Becoming "Human" a Solution?

Can granting Sophia human rights solve the problem? Within the legal system, robots can possess two legal statuses: as the object of rights and as the subject of rights.

The so-called "object of rights" refers to robots as the property of people, enjoying protection by law. Similar to a pet cat owned by an individual, it cannot be claimed by others. If someone wishes to enjoy the company of the cat for a few days, they must negotiate with the cat's owner, not take the cat without permission.

The so-called "subject of rights" refers to robots having the same rights and obligations as humans, and infringing upon robots is akin to infringing upon humans. If a cat possesses the status of a subject of rights, then one would need to request permission from the cat itself to pet it, and the notion of an owner would not apply—legally, someone could only be designated as the cat's guardian.

Granting robots the status of objects of rights poses no difficulty and is uncontroversial among experts. Robots as objects naturally do not have the problem of bearing legal responsibility. Mistakes made by robots would hold the owner, user, or manufacturer accountable, negating any legal responsibility on the part of the robot itself. It's analogous to someone taking an individual's cat. Holding the cat accountable is nonsensical. As an object, the cat lacks the capability to protect itself. Legal protection of the cat ultimately safeguards the owner's rights to the cat.

Conversely, if robots don't bear responsibility, nor do their owners or users, then third parties receiving services from robots would refuse those services. For instance, patients might be wary of undergoing surgery performed by a robot surgeon, rendering even the best surgical robots useless. This would, at least in this domain, hinder the development of artificial intelligence technology.

So, is it feasible to grant robots the status of human rights subjects, just like Sophia gaining citizenship?

What problems will this encounter? What legal dilemmas will the determination of non-human rights subjects face?

Does solving the robot responsibility problem in Sophia's way ultimately bring more convenience or more trouble?

1.7 Will Sophia Topple the "Dominoes" of Machines Becoming "Human"?

What is the legal basis for Sophia to obtain citizenship? There is no argument, and no argument is needed, because it is just a show. What needs to be considered now is, will Sophia become a kind of "precedent", providing a basis for her to enjoy a series of other rights, and providing a logical starting point for this "artificial species" to be on par with humans?

In judicial decisions, judges not only refer to laws, regulations, and normative documents issued by administrative agencies, but can also refer to previous precedents. Since existing precedents are considered to conform to the spirit and provisions of the law, cases similar to the precedents should naturally receive similar judgments. Some countries are accustomed to using "case law", based on this logic. The law is a rigorous logical system. Any "window" opened at any point, like autonomous vehicles hitting the street, once started, is beyond human control. Even though judges have the power of free evaluation, their room for maneuver is limited.

If Sophia represents a new world, the birth of a world with "duality" in intelligence, then with the birth of dual intelligence and a dual-subject society based on dual intelligence, the human legal system, especially the basic legal principles, will face a series of challenges. And once a dual legal system applicable to two types of intelligence becomes a reality, further problems naturally emerge:

What is the relationship between the two sets of laws?

Does the law of machines submit to human law, or vice versa?

If it is the former, what is the difference from the current situation where there is only human law?

If it is the latter, where is the position of humans?

If they are independent of each other, what if they conflict?

Is it necessary to set common principles above the two sets of laws to balance their relationship?

How is this set of common principles formulated?

Does the process of formulating these rules require the participation of both subjects?

If so, where does a previously non-participating machine gain the right to participate?

Will humans ultimately decide the fate of machines?

If so, will a dual legal system or even a dual world become a legal reality?

Such questions can continue until they fall into a real paradox: human thinking enters a state of contradiction where affirmation negates and denial affirms, eventually being logically asphyxiated by itself.

It seems that Sophia does not represent a solution to the problem, but rather triggers a "domino effect" that brings more problems once activated.

To truly solve the problem requires clearer thinking.

1.8 What is the Ultimate Goal of Human Development of Robots?

What is the ultimate goal of mankind in developing artificial intelligence? Is it merely to create a tool for ourselves, or to simulate God and create a new intelligent species?

In the development process of artificial intelligence technology, two obvious main lines can be seen: one emphasizes real application scenarios, from industrial robots to robot "doctors", undertaking work that humans are unwilling to do or not good at. The other line invades realms originally exclusive to humans, from chess, Go to Texas Hold'em, and recently Mahjong, all successively mastered by machines. From writing, painting to composing, and even filmmaking, these fields have also fallen to machines. Until the appearance of the "robot show" citizen Sophia, artificial intelligence technology is indeed rapidly moving towards approaching humans. The bipedal walking Atlas invented by Boston Dynamics can even do backflips that most ordinary people can't do!

If one were to inquire about the initial motivations of scientists, the reasons provided would undoubtedly be rational. Rather than aiming to have robots replace humans in creative endeavors, it's more about verifying AI technology and exploring the limits of machine intelligence. But who can guarantee that in the hearts of scientists, or at least some scientists, there is no intention to invent a humanoid, like God, to create machines "in their own image"?

If it's just about creating an effective tool, there's really no need to design machines to be so similar to humans, let alone surpass humans, or even elevate machines to the same height as humans, granting them legal subject status, like Sophia.

If it is to create a new intelligent species or even a superhuman species, then the relationship between this species and humans is not only complicated, but may also be entirely out of human control:

Just as humans evolved from apes, can today's apes determine the fate of humans?

Human-made laws can regulate the behavior of individuals, organizations, nations, and even international societies, but are they sufficient to regulate the behavior of superhuman species, if humans are uncertain about the level of intelligence this species can ultimately achieve?

Given that more and more experts believe scientists can create such a species, how can they regulate the new intelligent species with human legislation?

What's more absurd is, if the goal is really to curb the potential out-of-control phenomenon of robots, why not limit the development of robots in advance and not let it go this far?

At this point, a structural problem beyond the subjective will of scientists has emerged.

1.9 Is Binary Intelligence and Even Binary Legal System the Fate of Mankind?

Regardless of the goals humans have articulated, the factor that will truly determine the future direction of robots is the ultimate fate of all human creations: just as God could only create humans in his own image, humans can only create robots in their own image.

Biomimicry is a basic strategy used by humans in the manufacture of tools or machines, which is most clearly demonstrated in the "oar" invented in ancient China and still in use today. The "oar" is a tool used on boats. Its shape and operating principle are very similar to a fish tail, so it can combine the functions of a paddle and a rudder, achieving both propulsion and steering.

Of course, there are also technological inventions that show no trace of biomimicry, and the "bamboo dragonfly", a toy invented in ancient China, is a perfect example. The "bamboo dragonfly" is made of bamboo and consists of two parts: a thin bamboo stick and a thin bamboo slice. After processing and twisting, it looks like a rotor. A hole is punched in the middle, just the right size for the bamboo stick to be inserted. The bamboo slice can rotate horizontally with the bamboo stick. When it reaches a certain angular speed, it can axially detach. To play, you simply hold the bamboo stick between your palms and rub vigorously, causing the stick to rotate and drive the bamboo slice, generating lift and thus detaching and flying into the air. Then, under the dual forces of air resistance and gravity, it rotates and descends. The harder you use, the greater the initial angular speed, the higher the ascent, and the longer the flight duration. The "bamboo dragonfly" can be considered as the earliest "helicopter" of mankind, as no species in the world moves in this manner, making it a truly imaginative technological invention.

In biomimicry, humans find a plethora of models to imitate in various aspects because, in terms of bodily functions and their mechanisms, humans have their limitations compared to almost any other creature in nature, making learning from them nearly an inevitability. However, when it comes to intelligence, humans, considering themselves the pinnacle of creation, find almost no model for imitation beyond themselves. This is not to say other beings lack intelligence, but rather, the rationality shown in the means to achieve ends is fundamentally the same across all life. Thus, whether learning from ourselves or other species makes little substantial difference in this regard. However, considering the complexity of means-to-end logic in survival strategies, humans have reached the highest level of any species on this planet.

What's more rare is that humans can only recognize the intelligence of other species through observation, but they can understand their own intelligence not only through understanding, but also through the expression of the understood. Humans can not only express themselves, but also reflect on themselves, which is a basic prerequisite for some scientists to experiment on themselves. It is believed that what distinguishes humans from other species is not thinking itself, but "thinking about thinking", which is a unique ability of humans. Therefore, when creating a new type of intelligence, imitating human intelligence is the most suitable and effective strategy. In this regard, simulating humans is a choice without choice for humans!

According to the general logic of creation, even if humans create tools to serve themselves, these tools will inevitably surpass humans in at least one aspect. Therefore, the new form of intelligence created by humans, whether it will surpass humans themselves, and once it surpasses, whether it will lead to the birth of a new species, the so-called "silicon-based intelligence" relative to "carbon-based intelligence", is not a technical issue, nor is it a matter of will, but a natural logical issue. As scientists continue to develop artificial intelligence at the current pace, AI's realm of action is destined to evolve from certainty to uncertainty, from quantifiable to vague, from logical to emotional, from specialized to general, from unconscious to conscious, and from will-less to willful, advancing layer by layer in a technological evolution that is not diverted by human will!

Even with a clear research strategy, scientists' purposeful actions cannot completely prevent unexpected consequences!

Human motivation is not sufficient to influence the evolution of the universe. In the scientific context, this is a law. In traditional Chinese discourse, this is the way of heaven, which cannot be resisted!

The transition from wolves to dogs in the history of animal evolution provides an excellent case study of how new species of intelligence can emerge independently of human will. The development of artificial intelligence technology in the future might offer a contrasting example, revealing how purposeful human actions can lead to the emergence of a new form of intelligence that is unexpected and even contrary to human intentions.

1.10 The Birth of Dogs: Domestication or Evolution?

Dogs originated from wolves, a fact scientifically verified. However, the exact process of how wolves evolved into dogs remains debated. The conventional view is that humans domesticated wolves, turning them into tools for herding, hunting, searching, or even emotional companions. Yet, some biologists argue that the transformation from wolves to dogs was not due to human domestication but rather the dogs' own evolution.

1. Domestication theory

The domestication theory argues that early humans found newborn wolf cubs while hunting or in other situations, and for some reason, brought them back to raise. Like many animals, due to being isolated from their species and regulated by humans, after generations, wolves eventually turned into dogs.

Animal domestication should start early enough. Otherwise, it would be hard to change their wildness, especially for wolves. Zoological research shows that wolf cubs must be separated from their mother within 14 days of birth in order to be domesticated. After this point, the window for domestication automatically closes, and no human effort could domesticate a wolf.

Herein lies a question. In early human history, what was the purpose of raising a 14-day-old wolf cub? Did humans have the ability?

First, humans had no motivation to domesticate wolves. Wolves were never the main diet of ancient humans and, before becoming dogs, weren't considered as tools or objects of emotional attachment, leaving no rational motive for raising wolves. The only consideration is that raising wolf cubs was due to some primitive religious need, which needs to be verified.

Second, humans lacked the ability to domesticate wolves. A 14-day-old wolf cub cannot eat meat and needs to be fed milk. Where would the milk come from? It's unlikely that early humans had the means to produce milk suitable for wolf cubs.

Lastly, humans were neither willing nor capable of bearing the cost of domestication. With unstable food supply, especially protein, for themselves, feeding wolf cubs, even just with meat, would have been an undue burden.

The domestication theory is widely accepted because it satisfies human vanity as the "measure of all things" and conforms to the belief in human autonomy in arranging the world order. However, upon closer examination, there are indeed several missing links in the evidence chain.

2. Evolution theory

The opposing view is the evolution theory, which argues that wolves naturally underwent habit changes in the process of actively contacting humans, and eventually evolved into dogs. This process is much simpler than domestication. A pack of wolves, through genetic mutation, acquired the ability to digest carbohydrates, thus stumbling upon a "pie in the sky" opportunity. Simply by following humans, wolves could feed on human leftovers, unappealing or indigestible to humans. Since wolves don't care so much about the "taste on the tip of the tongue", as long as there is food, and it is more comfortable than hunting by themselves, why bother to travel long distances and often come up empty? Getting close to humans became the most reasonable choice for the wolf pack, although philosophically, humans do not recognize that wolves have the consciousness of choice. This change in habits eventually led to two major gene changes in wolves, and their directions were exactly opposite.

Firstly, the brutal genes were naturally eliminated. Wolves that attacked humans or too aggressively competed for human food were killed, interrupting the gene transmission. Over time, wolves that would not attack humans became dogs. To this day, some breeds of dogs still retain varying degrees of aggressiveness but with a clear direction: they only attack on their owner's command, not the owner themselves. Attacking the owner often results in direct genetic elimination.

Secondly, the cute genes were strengthened. To get close to humans and hope to get human favor, the best way is to show an acceptable cuteness. The cuter they were, the more food they got, the higher their chances of survival, and the higher the success rate of gene transmission. After generations of reinforcement, the fierce and ferocious wolves turned into the tail-wagging, pleading dogs.

Today, the dogs raised by humans can basically be divided into two categories, working dogs and pet dogs. The former retains some aggressive genes, while the latter almost exclusively has genes for cuteness.

The process of wolves becoming dogs may still need to be verified as to which path they took, but in terms of the smoothness and efficiency of the process evolution, the evolution theory is more in line with experience and logic than the domestication theory. The key difference between the two lies in the role of human effort in them.

In terms of the importance of dogs to humans, the transformation of wolves into dogs is definitely in line with human needs. However, this seemingly purposeful process might have nothing to do with human intentions. From the perspective of domestication, the transformation of wolves into dogs was intentionally directed by humans, who not only had ideas and utilized knowledge, but also took effective measures to force wolves to evolve step by step into today's dogs according to human needs. In the whole process, human intention played a leading role.

From the perspective of evolution, the transformation of wolves into dogs was not entirely independent of humans, as without the human application of both incentives and deterrents, the ferocious genes in wolves would not have been weeded out, nor would the genes for cuteness have been reinforced. However, without the initiative of wolves, nothing will happen, just like the wolf packs that still exist in nature today. The general laws of evolution, the plasticity of wolves, the new relationship pattern between specific wolf packs and humans, and the attitude of specific human groups towards wolves, together all jointly promoted the evolution of wolves into dogs. In this schema, human intention does play a role, but it is not as dominant as in the domestication schema. In other words, the process of wolves turning into dogs and dogs being bred into various breeds are not following the same evolutionary path!

The evolutionary process of wolves becoming dogs has already been completed, and whatever viewpoint zoologist hold cannot change this reality. However, the underlying principles offer much to ponder: processes of change beneficial to human purposes can occur independently of human intentions! This fact and its underlying logic pose a fundamental question for us to consider the future of artificial intelligence.

1.11 Where Does the "Autonomous Evolution" of Artificial Intelligence Lead?

Indeed, many scientists do not intend to create new intelligent species as their goal, and they admit that with today's theories and technologies, humans may not have the ability to achieve such a goal. Yet, throughout history and even today, it is not uncommon for humans to unwittingly head towards outcomes contrary to their intentions. In the development of artificial intelligence and its consequences, can humans escape this fate? Experts, now full of confidence in developing AI technology, have started advocating for "AI governance", influenced by various underlying psychological factors, including such concerns, perhaps.

Real life has repeatedly proven that, although humans proudly believe that the difference between them and other species lies in the fact that human actions are "purposeful actions", while animals only have instinctive behaviors, the results finally obtained by human actions may not necessarily be in line with expectations. The "unexpected consequences of purposeful actions" are normal in the sequence of human actions, not surprising but warrant caution. The question is, will the final outcome of AI technology development be an unforeseen, unbearable surprise for humans?

In fact, the pursuit of unexpected discoveries brought about by artificial intelligence has always been the internal driving force of this technology development. The final proof that machines can also think must be something that humans have never thought of, otherwise how to show that machines also have intelligence, how to prove the value of machine intelligence?

It's well-known that the development of artificial intelligence also undergoes an evolutionary process, starting from operating entirely based on the programming languages given by humans, so-called programs, to calculate and provide the results needed by humans. It first simulates human thinking that has been achieved, utilizing algorithms to learn from human experiences, such as chess games, and, driven by vast amounts of data, achieves goals set by humans with unimaginable computational power. Then, it simulates the evolutionary process of human thinking, learning on its own based on just the basic rules, constantly iterating, and evolving intelligence and outcomes beyond human imagination. Yet, this evolutionary process still unfolds along the objectives set by humans, so the results, while meeting specific scientific expectations, do not lead to unexpected discoveries.

The cutting edge of technology development now brings news of a new evolutionary pathway being explored at the edges of traditional evolutionary algorithms. Scientists no longer preset the goals that the algorithm needs to achieve, but let the algorithm evolve on its own, hoping to finally harvest unexpected and novel results. This evolutionary algorithm, adhering to the "stepping stone principle", is said to potentially grant artificial intelligence the capability for innovation, offering discoveries not anticipated by scientists!

Whether artificial intelligence can achieve innovation in the strict sense is a discussion for another time. The most anticipated and concerning aspect of this research strategy is: what kind of robots might the ultimate goalless evolution bring to humans if it becomes a common development strategy?

Can humans accept this "unexpected consequence"?

In nature, will it be a wolf turning into a dog, or a dog turning into a wolf?

Chapter 2
Is "Machine Will" a Contradiction in Terms?

Granting Sophia citizenship was merely a publicity stunt by corporations and governments, as AI technology has not yet evolved to the point where machines possess will, the ability to exercise rights, fulfill obligations, or bear responsibility. Thus, citizen Sophia exists in name only, lacking universal significance. However, isolating the question of whether machines can hold the status of rights subjects is not without value. From the development trend, whether artificial intelligence can have consciousness depends on how consciousness is defined. More likely, before artificial intelligence has consciousness, the relationship between machines and humans has already undergone major changes, and the issues of rights, obligations, and responsibilities caused by artificial intelligence are therefore inevitable.

Therefore, despite how absurd Citizen Sophia's emergence might seem, it still serves as a starting point for serious reflection. Legislation need not aggressively precede life, but legal thought must not lag behind it.

2.1 What is the Legal Significance of Free Will?

Semantically, free will initially emerged in discussions about whether humans possess divinity or religiosity, with human free will stemming from God. The so-called "subjectivity" issue involves human nature, related to human value, dignity, etc. The latter, focusing on issues post-Descartes, aims to escape divine influence and affirm human status. The later saying that "God is dead" indicates that human free will is self-sufficient and does not need to seek its origin in divinity.

Since the advent of the French Civil Code in 1804, civil law has particularly emphasized the autonomy of legal subjects. Specifically, a person is a subject in law, and all actions of a person are made according to free consciousness, not subject to coercion or fraud by anyone. Actions under coercion or force are generally considered revocable in law. Since actions are determined by one's free will, autonomy

J. Gu and C. Xu, *Will and Responsibility*, Human Intelligence,
https://doi.org/10.1007/978-981-97-3929-5_2

corresponds to personal responsibility, meaning one must bear legal responsibility for their actions.

Specifically, when an actor fails to fulfill a certain duty of care, resulting in harm, they must bear responsibility for such harm. But if we did not commit this act ourselves, then there is no need to be responsible for this act.

Legally, humans are first and foremost natural persons, endowed with free will and natural personalities. Under modern legal systems, individuals with natural personalities uniformly become legal subjects. Persons in law, through their actions of free will, bear responsibility. Here, free will is transformed into a legal system setting—the system of consciousness or behavioral ability, focusing on identifiable actions and their contents and consequences.

That is to say, law as a social norm, whether used to maintain rights, regulate behavior, or regulate interpersonal relationships, is based on the premise that people have free will. Knowing one's own actions or inactions, understanding why one acts or does not act, being aware of the consequences of those actions or inactions, and in such a situation, choosing to act or not to act is both a form of individual exercise of rights and the place where individuals bear responsibility.

Article 13 of the General Principles of the Civil Law of the People's Republic of China stipulates: "Natural persons have civil rights capacity from birth to death, enjoy civil rights and bear civil obligations according to law."

This provision clearly targets natural persons, meaning individuals with physical bodies in life, excluding robots or artificial intelligence.

The civil rights capacity of natural persons, simply put, is the status of rights subjects, legal personality, or the qualification to be a legal person in law. From birth to death, the subject qualification and rights capacity of a natural person will not change due to any event. No one can deprive a person of their legal personality, even if incarcerated. Only time can deprive a person of their legal personality. Upon natural death, a person's personality is extinguished in legal sense.

Article 14 of the General Principles of the Civil Law of the People's Republic of China stipulates: "The civil rights capacity of natural persons is equal." Whether born with mental retardation, disabilities, or engaged in criminal activities and punished by law, everyone's qualification as a person is the same because they are all of the same category of being, biologically the same. This equality is also the "ethicality of personality" in law or as Kant said. Since people are all the same, there will be no legal disparity between people.

2.2 Does Free Will Have Boundaries in Law?

On the surface, it appears that all humans should enjoy legal rights and bear legal responsibilities, but this is not entirely the case.

Without free will, even humans cannot exercise complete legal rights or bear the due legal responsibilities.

For example, courts often determine whether a person should bear legal responsibility by assessing whether they have the capacity to act, which is underpinned by free will and its realization.

When someone commits murder, defense attorneys, given the indisputable fact of the defendant's action, will seek conditions under which their client might not bear legal responsibility, such as mental abnormality. If a continuous abnormal state cannot be proven, it must at least be shown that the person was under an abnormal mental state at the moment of the murder. The concept of "intermittent mental disorder" as a defense strategy comes from this, aiming to deny the capacity to act and the underlying free will to evade legal responsibility for the related actions.

In essence, legal rights and responsibilities should be equivalent: as much right as one has, so much responsibility one bears, and vice versa. However, with "intermittent mental disorder" patients, a separation or misalignment of these two can occur. When normal, they are capable of exercising their rights, which others cannot deny. When not normal, they may fail to fulfill their responsibilities, which others cannot demand. This results in a peculiarly advantageous legal status, akin to a privilege, where one has rights without responsibilities.

The frequent use of "intermittent mental disorder" as a defense by defendants and their lawyers is not only legally based but also philosophically logical. A person always in a state of mental disorder cannot self-prove their condition. Being able to reasonably assert one's mental disorder essentially proves their mental competence. Intermittent mental disorder allows for the claim of incompetence during specific periods without fearing that one's behavior negates one's claims.

At the moment of implementing a certain behavior, even a sound person may not need to bear the legal responsibility of the behavior because they cannot exercise free will. For example, "Soldiers take obedience to orders as their duty", and ordinary soldiers are not legally pursued for executing orders because, despite having the capacity to act, obeying orders is a duty without choice. This is not only because disobedience will lead to being "subject to military law", but also because without executing orders, the army, law, and even the country cannot exist. The obligation of "must execute orders" must be accompanied by "no need to bear the responsibility of executing orders".

The law cannot require individuals to sacrifice themselves to preserve others, which is reflected in situations such as justifiable defense and emergency avoidance. Because every person and their free will are equivalent, there is no difference.

Those executing orders are not subject to legal prosecution, but nuances existed in detail. After the unification of East and West Germany, judges tried former East German soldiers who shot civilians. When defense attorneys argued that soldiers were just following orders, counterarguments were made that soldiers could have "raised the gun barrel by three centimeters" when firing. This means that following orders doesn't entirely eliminate flexibility. An individual could use all opportunities and conditions to avoid severe consequences but chose not to, indicating subjective inaction. Therefore, after considering the special circumstances of executing orders, the legal responsibility for rigidly executing orders still needs to be pursued.

A real case from a few years ago involved several people kidnapped, and one of the male captives was forced to rape a female captive under the threat of death. He complied. After being rescued, he was charged with rape.

Can legal forgiveness be granted for violating others to save oneself under coercion?

This depends on the ethical standpoint.

One viewpoint believes that raping others out of coercion is a survival behavior and can be exempted from responsibility.

Another viewpoint holds that, despite being forced, any sexual act against the other's will constitutes rape.

The fundamental difference between the two viewpoints lies in whether two independent free wills are equal. Preserving one's own life is a requirement of free will, and everyone's life is precious. However, one cannot deem another's life less valuable or another's free will subordinate to one's own. If one believes their life is more valuable, can they, during an air raid with insufficient bomb shelter space, expel others to save themselves? Why should one person's free will to seek shelter and save their life be subordinate to another's?

Because humans possess free will, they must bear the legal consequences of their actions. Once free will is lost, even if one has the ability to act, one is not within the scope of legal prosecution. Of course, the loss of free will not only has conditions but also degrees, and the extent of legal responsibility must match the degree of lost free will.

2.3 Certainty or Uncertainty, How Does the Brain Position Itself?

In fact, whether people have free will and how free will is generated is still controversial. However, many concepts, serving as tools, may not derive their value from themselves but from the subsequent discussions they provoke. These concepts are considered to have methodological value.

Free will serves as an analytical tool. Therefore, aside from the doubts about free will, this concept can still be used to spark discussions and yield valuable results. It's important to remember that all reasoning is conducted with this concept as a logical starting point.

Famous psychologist John Searle once said that all our consciousness, including the process of consciously making decisions, completely depends on lower-level brain processes. Biologically speaking, the activity of neurons in the brain determines human consciousness. On the basis of consciousness, the ability to make free choices can be formed, and actions can be chosen according to one's needs. Intentional consciousness achieves free will, and only free will can give actions a variety of possibilities. The fundamental difference between human actions and object movements lies in whether they have voluntary intentions. Without consciousness, without the intention of free choice, there can be no real action.

Then comes the question. Based on knowledge about the rules of neuronal activity and the current state of the brain, is it possible to predict the future operation of the brain? This leads to two different views: determinism and indeterminism.

The deterministic viewpoint advocates that every system has basic rules and causal connections. When applied to human society, all human activities are the result of some previous cause or causes, and human behavior can be accurately predicted based on previous conditions and experiences.

Indeterminism, on the other hand, believes that it is impossible to obtain knowledge of all rules and information about all states, and that the operation of rules is conditional and even variable, with the state itself being unstable. Therefore, it is impossible to predict based on rules and states.

The saying by ancient Greek philosophers that "one cannot step into the same river twice", the uncertainty principle and quantum state collapse in quantum mechanics, and the breakthroughs in psychology that claim psychological theories become ineffective because people who master these theories change their behaviors, all belong to the indeterminist camp. Using the words of Confucius to highlight the difference between the two, the question is whether humans can "do what is known to be impossible"?

We both acknowledge that "existence determines consciousness" - human ideas are not without basis, agree that "consciousness can act upon existence", changing it to alter consciousness, and do not avoid the fact that in the same existence, different people can have different consciousness.

Such thinking is open enough to incorporate cultural diversity into the theoretical framework, and can also raise new valuable questions: if the same existence can lead to different consciousness in different nations, then what is the internal mechanism by which existence determines consciousness?

Whether the brain is a deterministic system is controversial, but the law does not allow such controversy. If the human brain belongs to a deterministic system, talking about human free will is a pseudo-proposition. How does a determined person possess freedom, will, let alone free will? Only by positioning the brain as a non-deterministic system can the free will in a practical or experiential sense have room to exist.

However, the subsequent neuroscience question is, how to study the laws of a brain that is not determined by external factors or internal logic? Different individual brains have biologically similar structures, such as neurons and circuits, and have common properties and operating modes, such as bioelectric currents, but the results are different. How are these differences formed? Scientists face individual brains in their research, but the findings need to apply to all normal brains and even to a diseased brain, its pathological discoveries also need to apply to other brains showing the same symptoms. The universality of knowledge, which is opposed to human individuality, requires commonality or similarity as a bridge to cross the two. Where does this come from?

When a problem is insoluble in philosophy, humans wisely turn to practice for solutions. Whether free will is an illusion or an empirical fact does not need to be tangled at the philosophical level and can be discovered through experiments, observation, and science. Unfortunately, to date, human understanding of the brain is too poor and shallow to provide a final conclusion.

Nonetheless, following the path of empirical inquiry, further analysis of free will can still be conducted. In fact, a clue for deeper interpretation is hidden in John Searle's statement.

Based on consciousness, humans can make free choices. With consciousness, one can choose to act, not to act, or how to act. For example, they are free to choose whom to contract with or which brand of mobile phone to buy. Only when humans have chosen actions, with intentional consciousness, can free will be represented. This is the relationship between consciousness and will. Conscious behaviors and unconscious actions are different. Things like sleepwalking, conditioned reflexes, and subconscious actions, such as meaningless gestures during a speech, body shaking when nervous, etc., cannot be called behaviors, but only actions. In a biological sense, the relationship between consciousness and free will is clear.

In a nutshell, humans can possess free will in terms of empirical facts. As practical subjects, the essential attribute of humans is free will, which also forms the basis of moral behavior. Since free will as a philosophical concept emphasizes causality, i.e., what it is based on, not how it arises, it is appropriate to say that vegetative people also have free will. Even though the brain is dead, as long as the law does not accept the definition of "brain death", vegetative people are still considered to have all the rights that normal people should enjoy, which only need to be exercised by their guardians.

The dispute over the positioning of the human brain has not affected the continuous role of the brain in scientific research, but in the legal field, it is related to the operation and even the existence of the law, as the law adjusts conscious behaviors, and the view of the brain as a deterministic system is akin to fatalism. To extreme fatalists, everything they do is driven by fate, not the result of autonomous choices. Since there is no choice, there is naturally no responsibility, so why should the law pursue it?

For its own existence and value, the law cannot but reject the view that the brain is a deterministic system.

2.4 Is Human Free Will Merely a Presumption?

In jurisprudence, human free will possesses an a priori nature and can be regarded as a presupposition of legal reasoning. To put it extremely, a person without free will is not a human. The saying "Give me liberty, or give me death" captures this essence.

If one seeks a more tangible basis for human free will, it can be found both in biology and culturology.

1. Biology: humans are animals of uncertainty

Proving human free will from a biological perspective is relatively simple. In a nutshell, free will is the uncertainty that humans set against nature. Unlike other animals that strictly adhere to the natural order, humans can "legislate" for themselves and live in a "man-made nature".

Among all living beings, humans are perhaps the least naturally endowed: not the fastest runners, incapable of flying, born without the ability to swim, lacking fur for protection, and without sharp claws. If not for their intellectual and manual dexterity, humans would likely have been the first species to be eliminated by natural selection.

This intellectual and manual dexterity, the most uncertain endowment from nature, is what allows humans to continuously improve themselves, adapt to and even change nature, and create things that neither exist in nature nor could be produced by it, including humans themselves, distinct from apes. The reason why humans can improve themselves and change the world lies in their uncertainty. Because of uncertainty, humans can choose, and choice is the proof of the existence of free will.

Interestingly, too many choices can become a burden for individuals. In daily life, responding with "whatever" when asked for a preference is often seen as a virtue of modesty. But in law, not choosing is also a choice. "Whatever" as a genuine expression of intention is also a way of exercising free will with legal effects and corresponding responsibilities, if any. "Avoiding choice" is also a choice, and "escaping freedom" is also a freedom that only free individuals possess.

In August 2019, the American retail company "Costco" opened in Shanghai, attracting many citizens, including consumers from other provinces and cities. One of Costco's business strategies is to allow consumers to give up the freedom to choose brands in an era of too many brands. Although there is consumer trust in the store behind the abandonment of choice, it is not too arbitrary to say that the members of this company have an inherent tendency to give up choice.

Free will manifests as choice, choice comes from uncertainty, and uncertainty is the characteristic of being human, the arena where will is exercised!

2. Culturology: humans are symbolic animals

In culturology, uncertainty is more of a physiological characteristic of humans, rather it is the essence of their intelligence.

Philosophers like to propose their own definitions of human nature, consciously or unconsciously using other species as reference objects.

Aristotle said, man is a political animal.

Bacon said, man is a tool-making animal.

Marx said, man is a labor-producing animal.

These definitions can stand within their respective domains but also have limitations. In other animals, there are behaviors that are not substantially different from the above definitions, unless these behaviors are stipulated as unique to humans, it is difficult to limit these defined behaviors to humans only. However, in doing so, one inevitably falls into the logical dilemma of circular definition.

To say that animals have no politics, the hyenas under the rule of the "queen" would certainly disagree.

To say that animals can not make tools, the handy gorillas would certainly disagree.

To say that animals do not engage in productive labor, the leaf-cutter ants that live by cultivating fungi would certainly disagree.

The final way humans maintain their dignity is to deliberately emphasize the intrinsic differences in behaviors shared with animals: in animals, it's instinct; in humans, it's conscious awareness or free will.

Modern scholars like German philosopher and anthropologist Ernst Cassirer (1874–1945) and American anthropologist Leslie White (1900–1975) believed that "humans are symbolic animals". Israeli author Yuval Harari, in his book "A Brief History of Humankind", proposed that the characteristic of human language is to "fabricate stories", which adopted the same research approach as the two predecessors. In their view, humans are symbolic animals, and concepts are also a type of symbol. Symbols require a physical carrier, but the meaning represented by symbols is not determined by the objective attributes of the carrier, but is completely designated by humans, with arbitrariness or autocracy.

An apple is a fruit that can be eaten for its crisp, sweet taste and rich nutrition. No matter how people view apples, under normal circumstances, these objective attributes are constant. The apple will not turn sour because someone who can't eat it feels sour. But an apple that has been bitten is different.

The famous "bitten apple" used by Apple Inc., the inventor of the smartphone, as its corporate logo, can be understood as the "fruit of wisdom" that opened the eyes of human ancestors to discern good and evil in the Garden of Eden, or it can be "misunderstood" as the poisoned apple bitten by English scientist Alan Turing, the "father of artificial intelligence", to escape unbearable legal punishment. Both interpretations are said not to be the original intention of Apple Inc., but in any case, attributing admiration for technological wisdom or remembrance of great scientists to this ordinary fruit is arbitrary or autocratic. With sufficiently creative interpretations, substituting the fruit with a pear or peach would be equally feasible.

Defining humans as symbolic animals is intended to reveal the essence of humans and their free will. Unlike animals confined by natural boundaries, humans can break through natural limitations and achieve self-design. Like the nature of symbols, humans face no boundaries when making settings. Any object can be assigned any meaning, and even constellations, over which humans have no influence, can be endowed with special significance. The Big Dipper's use in navigation is a function brought by its natural properties, but it could not have inspired human emotions before it became a symbol. Because humans always assign meaning to everything they perceive, they essentially live in a world of symbols, an uncertain world.

In recent years, young people in China seeking a "sense of ritual" in everyday life essentially perform an operation of symbolizing ordinary activities by endowing them with extraordinary meanings, briefly termed "symbolization of life".

Symbols have arbitrariness or autocracy, and people living in the world of symbols naturally become a kind of "uncertain animal".

The most difficult barrier for machines to simulate human intelligence is not the perception and processing of deterministic entities, but the grasp of uncertainty. Sensors can accept real information, extensive enough to surpass the corresponding human organs, but can they also sense the meaning of symbols? Without prior knowledge, can exhaustive physical and chemical characteristics of apples allow machines to understand the logo of Apple Inc.? At most, they can compile a "literature review" of all interpretations regarding the bitten apple.

2.5 Do Humans Truly Possess Free Will?

In the future, when super artificial intelligence breaks through the door of the human symbolic world, machines will definitely give this definition when mocking humans: humans are a paradoxical "carbon-based existence" that bustles between the dichotomy of certainty and uncertainty. Humans, because of their inherent uncertainty, are forever in pursuit of certainty. From religion to science, from fortune-telling to algorithms, they are all asking one question: can the world be certain? They are all chasing one goal: to make the world certain!

Confucius believed in "knowing that it is impossible and doing it", revealing the potential of humans to break through natural causality-the best annotation of human uncertainty. Yet, he also states, "If I hear the Way in the morning, I can die content in the evening", meaning that as long as the ultimate truth is grasped, one's wishes are fulfilled and one can die without regret. There is no big problem with this intention itself, but when placed in the contradiction between certainty and uncertainty, the absurdity is revealed.

How can humans, as uncertain beings, make the world certain as their mission or the completion of their mission? Have they ever thought about how uncertain beings will cope once the world is determined? "Hearing in the morning and dying in the evening" may be the only logical exit to escape this paradox: the day when uncertainty is determined is the time when uncertain beings do not exist.

Perhaps anticipating this paradox, Laozi, who predates Confucius, advocated that "the Tao which can be spoken is not the eternal Tao". Any "Tao" that can be articulated is not the Tao itself, so for any statement about the Tao, listen but reserve belief. Why trouble yourself by taking it as absolute truth or even letting it drive you to extremes? Allow the nature of uncertainty to exist as it is.

Indeed, the tension between certainty and uncertainty reflected in humanity is ubiquitous, requiring only our careful and sharp observation.

For example, suicide occurs daily in the human world, met with sympathy but also acceptance as commonplace. Yet, the notion of "mass animal suicide" strikes people as bizarre, underlying an unspoken consensus: suicide, as a defiance of the genetic imperative, should only occur among humans with the free will to break the causality, not animals strictly adhering to natural laws, otherwise it is equivalent to admitting that animals also have free will.

Perhaps, the so-called only humans have free will is a metaphysical autocracy, reflecting the vanity and even presumption of anthropocentrism, and animals may not lack free will. Faced with a dog that starves itself in mourning, no one will doubt its understanding of the consequences of its actions.

Perhaps, humans lack understanding of the spiritual world and behavioral logic of animals, and the term "collective suicide" itself is misleading. In human society, the deceased in a stampede accident would definitely protest with their last breath if they knew they might be misinterpreted as "collective suicide".

Regardless, maintaining the unreflected monopoly of humans on the right to suicide certainly concerns the dignity of human free will.

What destroys the dignity of human free will is statistics, which are indispensable to artificial intelligence!

Normally, suicide manifests as an individual behavior, with each person who commits suicide having their own reasons, including some unthinkable thoughts, fully reflecting the freedom of personal will on the matter of life and death. Strangely, in any society without significant changes, the suicide rate of a specific group appears quite stable, with fluctuations even lower than the death rate due to illness.

This brings intrigue.

Death from illness falls within the realm of natural causality. The struggle of mentally healthy individuals to exhaust all medical means to 'cling to life' drives the continuous development of medicine. Conversely, those who seek illness without being sick and end up dying can at least be judged to be mentally ill. Suicide is a personal decision and falls within the realm of free will. Each suicide victim has their own reasons. When they step onto the window sill out of their own decision, they neither care nor know about the suicide rate of the year, let alone persist in inquiring "how well the target is completed this year", until they confirm that there is a balance before they jump. In this way, isn't it strange that the suicide rate remains surprisingly stable?

The simplest logical explanation for this is either that all individuals' free wills cancel each other out, ultimately manifesting as a lack of collective human will, or that behind each suicide victim there exists a mysterious force that influences a certain number of individuals each year to passively approach death under the guise of exercising free will. In this case, for those who are in the same crisis but haven't committed suicide, is their restraint due to a strong nervous system and firm will, or simply because they haven't received a directive to act and thus remained inactive, rather than deliberately choosing not to act? Which is the more realistic possibility?

In fact, as early as the nineteenth century, Belgian statistician Lambert Quetelet (1796–1874) discovered that in a society with a basically stable state, the number and proportion of crimes committed each year, and even the number and proportion of different types of crimes, have the same stability as the suicide rate, and proposed the concept of the "average man". The "average man" represents the statistical abstraction of human group characteristics, not an entity showing individuality and will, akin to average height, income, lifespan, etc.

While crime does not directly challenge the natural law of genetic transmission like suicide, committing a heinous crime with severe consequences that ultimately leads to a death sentence has the same biological effect as suicide. Does this imply the same mysterious force behind capital criminals, driving individuals towards death, whether by suicide or homicide, leading to the same end?

If, in the microcosm of suicide, the certainty and uncertainty of humans are manifested as the tension between the uncertainty of the individual and the certainty of the collective, then in the relationship between humans and culture, the tension between the certainty and uncertainty that is also represented is that the collective has uncertainty, while the individual has certainty.

Culture, created by humans, bears the relative uncertainty of nature, vibrant across different human cultures. Conversely, individuals are determined within their culture. Culture manifests as a common value concept and behavior mode of a nation, not subject to individual will. Festivals are scenes where all members of a nation act uniformly under cultural influence. If any ritual entails self-harm or suicide characteristics, could it be considered, like suicide, to follow some directive?

In fact, the concept of social science itself implies that human individuals possess free will, yet the human collective does not show free will, rendering "social laws" moot without this presupposition. Whether it is the dispute between "social nominalism" and "social realism" in sociology, and the "methodological individualism" as a compromise between the two, the historical debate of "times make heroes" versus "heroes make the times", or the tension between rational calculating individuals and the "invisible hand" in the field of economics, they all express the same meaning: how individual uncertainty eventually dissolves into collective certainty of humans!

The existence of supernatural mystical forces remains unannounced, yet the domination of human will by machines has increasingly become a serious issue. In the internet environment, artificial intelligence combined with big data has created a scenario where algorithms control personal information. Merely leaving online data, whether clicking on news or gossip, making payments or receiving them, producing or consuming, or even just appearing without clicking, will be "recorded in history" by facial recognition and then subtly guided by machines. For those highly dependent on information technology, "machines know them better than they know themselves", with machines guiding humans and even deceiving them, has become a reality.

Whether humans should listen to machines or machines should listen to humans poses a question.

Even if it has not become a problem for the human collective, it certainly poses a serious issue for a significant number of individuals!

Such speculative thoughts, while intellectually entertaining, once they become a part of real-life scenarios, are sure to send shivers down one's spine. A legal issue with both theoretical value and practical significance arises unexpectedly:

If there is a mysterious instruction behind human behavior, whether it comes from nature, culture, or machines, can humans still claim to have free will and should they be responsible for their own actions?

2.6 Does Artificial Intelligence Have Free Will?

Anyone who has some understanding of the current state of artificial intelligence research will not hesitate to answer: it is too early to talk about artificial intelligence having its own consciousness or will.

Although AlphaGo has triumphed over top human Go players, it subtly reveals the shadow of a programmer with the skill of a 6-dan Go player. At present, artificial intelligence remains in the "weak artificial intelligence" stage, surpassing humans in single tasks but incapable of managing various jobs with one brain, as humans do.

The Western proverb goes, "For the average person, doing one thing is one thing, for the smart person, doing one thing equals several things, and for the genius, doing one thing equals everything!".

Similarly, ancient Chinese expressed this idea with four characters: "融会贯通" (integrate and understand) or "举一反三" (infer from one instance).

With the current level of technology, artificial intelligence can only complete specific tasks in a single field where intelligence is needed under human settings, such as playing chess, reading images, and text analysis. Although its efficiency is far higher than that of humans, it becomes clumsy once it steps out of its designated area. Don't even mention asking a Go-playing AI to drive a car or play Mahjong, as Go is a complete information game, while Mahjong is an incomplete information game. Humans can easily switch between these games, adapting with a simple change in rules, a flexibility AI cannot achieve without a complete overhaul.

Upon closer inspection, many times what appears as "outstanding performance" by artificial intelligence is achieved through extremely basic methods that would be considered "stubborn" if done by humans. The perceived cleverness of artificial intelligence, mistaken for surpassing human intelligence, is merely due to its vastly superior computational power.

Herbert Alexander Simon, a renowned AI expert, once provided an easy-to-understand and amusing example to illustrate the difference between human thinking and artificial intelligence.

The equation "DONALD + GERALD = ROBERT", where letters stand for numbers, contains ten different letters representing the numbers 0–9. The task is to figure out which number each letter represents so that the addition of two words equals the third, starting with the condition that $D = 5$.

Faced with this problem, artificial intelligence excels at brute-forcing through all possible combinations, trying every potential numerical representation for each letter to see which works, considering over three million possibilities to find the solution. This "low-level thinking" is made possible solely by its vast computational power, allowing artificial intelligence to swiftly submit the work and achieve high scores.

"Outsmarting is better than overpowering." Humans lack such computational skills and will not blindly "work hard", allowing a game of little mathematical difficulty to consume their lifetime. Humans approach it much more efficiently. Because $D = 5$, $5 + 5 = 10$, so $T = 0$. Carry one, $1 + 2 \times L = R$, R must be an odd number. $D + G = R$, R is an odd number. Because it has been judged that $T = 0$, then R must be an odd number greater than 5, namely 7 or 9. O + E = O, in the case where there is no carry in the previous digit, E = 0, in the case of carry, E = 9. Since T = 0, only E = 9, R = 7 can be concluded. In the same way, it can be deduced that G = 1. By the same token, the numbers represented by the remaining letters are finally obtained, L = 8, A = 4, N = 6, B = 3, O = 2. Substitute one by one into the original formula DONALD + GERALD = ROBERT, and get 526,485 + 197,485 = 723,970, the equation holds true.

Humans devise strategies, utilizing set conditions and logical reasoning to find shortcuts and reach solutions. Machines, clearly, do not possess such wisdom yet and can only calculate all possible paths. However, to be fair, the lack of cleverness in artificial intelligence is also due to humans' insufficient wisdom to equip it with smarter algorithms.

In fact, the widely used Gaode Map employs a brute-force method. The so-called "Monte Carlo decision tree" lists all possible routes from start to destination, pruning less efficient branches, leaving several short or time-saving paths for the driver to choose from.

These examples show that the current "creativity" of artificial intelligence mainly relies on computation or statistics, operating within the rules and data provided by humans or collected on its own, and conducting massive calculations to produce the desired conclusions. That is to say, the so-called "creativity of artificial intelligence" is merely an illusion. Achievements that seem to match or exceed human intelligence are the result of rules and data specifically fed into AI systems by humans.

Although artificial intelligence has achieved self-evolving algorithms, capable of generating new algorithms through existing ones, as statistician Nate Silver remarked, "We might not describe the unique moves made by computers as creative actions, as their capability to do so may merely be due to their vast computing power." In this process, machines and systems merely recreate human intelligence or decisions made by humans. Even if AI's decisions surpass human levels in recreating human decisions, what artificial intelligence can currently do remains a copy of human determinations. Therefore, at least for now, artificial intelligence is unlikely to seriously claim, like Sophia, "I do not possess self-awareness."

2.7 What is the Practical Significance of Discussing Machine Will?

Raising the question of machine will, knowing full well that artificial intelligence does not yet possess will, begs the question: Why? "Mentioning what is not to be mentioned", is it just for sensationalism and to attract attention?

Not at all.

The history of human development of machines can be impartially described as the process of tool automation to meet human needs, and the historical process of automation can be described as the evolution of humans from seeking certainty in machine performance to embracing uncertainty.

The overall trend of automation is to continuously reduce human direct control over machines, both prior to or during operation. Simple physical or chemical mechanisms can achieve a degree of automatic machine operation. The combination of the inertial rotation of the flywheel and the explosion of atomized gasoline makes the car run automatically. At this stage, the emphasis is on the machine's certainty or reliability, demanding precise execution of specific actions as set by humans without surprises. Any performance deviating from the designer's expectation is deemed a mistake or unreliable. Patterned behavior is the hallmark of machines at this stage, where "mechanical" describes both the general category of machines and their rigid, inflexible nature.

The inclusion of information technology has elevated automation to a new level. Humans input signals to adjust machine actions. Machine behavior begins to exhibit variability and even plasticity, greatly enhancing adaptability and execution efficiency for different tasks. Yet, at this stage, machine behavior remains under absolute human control. Programming dictates machine actions, and unexpected outcomes are still not the hallmark of a good machine.

In the era of artificial intelligence, machines have truly become autonomous. Technicians simply set the goals, and the machines can achieve the results on their own, from data acquisition, rule application, path selection, error judgment, to decision execution. All these steps, while dependent on human-designed algorithms and pre-set programs and requiring vast datasets for machine learning, leave the execution process entirely to the machine, including handling unexpected situations due to environmental changes. By this stage, the measure of technological prowess is not so much the machine achieving the expected results, but rather the process of artificial intelligence achieving goals unexpectedly!

What astonishes people in AI's Go games or Mahjong matches is not the record but the strategy and method of achieving it. Tactics and strategies that humans never considered, even those long deemed off-limits, are overturned by AI, proving new optimal strategies in defeating human opponents. AlphaGo overturned the Go theory of "golden corner, silver edge, grass belly" accumulated by humans for thousands of years, introducing what top players saw for the first time as the genuine "golden belly". After the defeat, Ke Jie lamented: "I am human, AlphaGo is God!" as he, who was full of confidence before the game, never anticipated the machine could execute such sophisticated moves.

"When humans think, God laughs." It is God's usual style to be unexpected.

At the 2019 World Artificial Intelligence Conference, Microsoft announced that the Microsoft Asia Research Institute has successfully developed a highly skilled artificial intelligence system for Mahjong-"Super Phoenix" (Suphx). It is reported that this artificial intelligence has been promoted to a ten-dan player on the Japanese online Mahjong competition platform "Tenhou", reaching the peak level, by playing against real players. Professional Mahjong players from Tenhou were astonished after watching Suphx playing 5000 hands, noting its unconventional yet effective strategies.

From a human perspective, Go is considered more complex than Mahjong, which partially explains why there are more Mahjong players than Go players. However, for artificial intelligence, which relies on complete information, Mahjong presents a greater challenge. In Go, every move and its potential consequences are transparent to both players. In contrast, Mahjong players can only see their own tiles and those played by the other three parties, roughly analyzing the tiles that the other three parties may hold. As for the tiles that one might get later, at most a probabilistic inference can be made, and it is impossible to know whether a certain tile is still in the deck, let alone whether it will be obtained as desired, because it is impossible to determine whether the actual tile obtained will be completely different due to the "eat" or "touch" of the previous player. Making decisions based on extremely limited and incomplete information is the challenge that Mahjong brings to artificial intelligence far beyond Go. Mahjong requires not only skill but also luck, not only reasoning but also intuition. Now the question is, isn't it surprising that artificial intelligence has mastered Mahjong and has played unexpectedly effective hands?

In fact, for artificial intelligence, such as self-driving, isn't the greatest hope of humans that machines can automatically take corresponding measures in unexpected situations that humans have not thought of, and succeed?

Stepping beyond mechanically responding to tasks through pre-programmed human scripts, and achieving what humans have not thought, done, or achieved, is the true essence of artificial intelligence. Whether or not the prefix "artificial" is used is not a substantive issue.

Indeed, human expectations for machines are rooted in the standard demands placed on their fellow beings. Decision-makers require implementers to execute orders flawlessly while also demanding creativity in execution. "A general in the field does not follow the orders from the sovereign", and if machines truly possess will, they should also exhibit such responsibility and ability, right?

Empowering machines with the ability to choose is the core issue of artificial intelligence research.

In this sense, the "algorithm black box" should not be regarded as a defect in technological development, but rather as a new height reached by artificial intelligence: the invention of technology has entered a new phase, with humans beginning to lose grasp of their own creations.

The highly acclaimed creativity and innovation of recent years inherently demonstrate that groundbreaking knowledge, divergent from the existing knowledge system, is uncertain, not derivable from it, nor necessarily explainable within it.

Indeed, artificial intelligence, with computation or statistics as its "mode of thinking", follows a developmental trajectory from certainty to uncertainty. The leap from playing Go under full-information conditions to playing Mahjong under incomplete-information conditions reflects this trend and its amplification.

Uncertainty, the ability to choose, and volitional intelligence mark the coming of age for artificial intelligence!

2.8 Is Machine Will the Goal of AI Research and Development?

Admittedly, today, the notion of a machine possessing a will might seem as far-fetched as the idea of a machine thinking seemed in the past. However, given the achievements of artificial intelligence technology today, where machines thinking has become a reality, the possibility of machines having a will in the future cannot be dismissed. Perhaps then, the concept of machine will might no longer be considered a contradiction in terms.

In fact, the concept of machine intelligence proposed by the "father of artificial intelligence", British scientist Turing, includes such an intention. Simulating human intelligence has always been the basic strategy of artificial intelligence research. Since will is an indispensable part of human intelligence, machine will should also be considered an essential topic for discussion.

In recent years, artificial intelligence research has made significant strides, with its developments and applications far surpassing Turing's era. Yet, the philosophical foundation remains the Turing machine. When Turing set out to design machine intelligence, he established two goals: skill and mind, both of which remain valid today. Some people interpret this as, "Let the machine do meaningful things that usually require intelligence to complete, and simulate the process that occurs in mind based on physiology."

If the goal is to replicate, then within the scope of intelligence, anything human should be achievable by artificial intelligence. Humans have senses, artificial intelligence has sensors; humans have language, artificial intelligence can process natural language; humans understand the concept of seeking benefits and avoiding harm, artificial intelligence algorithms have rewards and punishments; humans have consciousness and will, how can artificial intelligence be without them?

Technological capability permitting, humanity will undoubtedly strive to emulate human traits in artificial intelligence as closely as possible, aiming to surpass human capabilities. This path of technology development has been proven time and again. The attempts made by scientists in the direction of biomimetic AI far exceed those in any other technological field. If intelligence is the crown of human self-replication, then consciousness and will are the crown diamonds. Today, the crown has taken shape, will the position of the diamond be vacant for a long time?

To take a step back, from a philosophical perspective, there may also be another prospect. Despite successful emulation, given that artificial and human intelligence fundamentally represent two different forms of intelligence, it's conceivable that while humans cannot do without consciousness and will, AI, with its powerful data gathering, computational capabilities, networked operations, and especially its ability to iterate on its algorithms, might discover new knowledge and construct new worlds, even without consciousness or will.

What sounds like science fiction today might not be so far-fetched after all.

2.9 Technically, Does Artificial Intelligence Have Free Will?

There is no straightforward answer to this question, because it not only involves whether it is theoretically feasible or technically feasible, but also regards whether humans are willing. Setting aside willingness here, if it's not feasible, discussing willingness is meaningless. "Being overenthusiastic alone" won't yield results. Conversely, if a new form of intelligence is part of the universe's natural evolution, human willingness or reluctance is inconsequential.

Experts in the field have varying attitudes but can generally be categorized into extremely optimistic, very optimistic, and less optimistic.

1. Extremely optimistic

Those with an extremely optimistic view are highly cautious about AI technology, not perceiving artificial intelligence as a future adversary to humanity, and thus not issuing early warnings about human order being overturned while machine will is still unformed. Influential figures like Hawking and Musk fall into this category.

Hawking warned many times before his death, "We are not sure whether we will receive infinite assistance from artificial intelligence, or be infinitely marginalized, even annihilated." He believed, "There's no fundamental difference in what biological brains and computers can achieve."

Clearly, Hawking recognized the differences between human brains and computers but noted that their outputs were becoming indistinguishably similar. Let's put it this way, assuming the brain is equivalent to a cow, producing beef, and the computer is equivalent to cloning beef cells, proliferating to produce artificial beef. Although the ways of producing meat are different, the beef cells and their nutritional components are the same. They are all proteins, amino acids, etc. From the perspective of ensuring that consumers maintain normal survival, substituting artificial beef for real beef is plausible and feasible.

Since it is feasible to replace beef with artificial beef, is it also possible to replace cows with petri dishes? Factory production of milk, rather than raising cows, is probably not good news for cows.

However, whether this will become a reality and whether cows will become extinct depend on what people need.

If it's just about nutrition, there's no problem. Taste can be accommodated, saving money is natural, and it might even be more eco-friendly. But for appreciating bullfighting in Spain, petri dishes are useless, even if cloning could produce a ton of bull-shaped artificial meat, it wouldn't be suitable for the bullfighting arena.

2. Very optimistic

If extreme optimism carries a style of science fiction imagination, then very optimistic views are more technologically styled.

The connectionist school in the field of artificial intelligence research can be classified into the optimistic camp, and the currently popular deep learning belongs to this school. The core view of this school is, "Neural networks and deep learning, by imitating the human nervous system, present the model of the human nervous system in the way of a computer, and use it to imitate intelligence."

The technical reason that allows these experts to be optimistic is that although humans do not yet understand what the essence of neurons in their brains is and how they are constituted, this does not prevent scientists from understanding what the structure of neurons is. Importantly, the lack of understanding of the nature of neurons does not hinder scientists from achieving intelligence by simulating the structure of neurons.

The most pure connectionists believe that it's not even necessary to understand human neurons, which are complex with over a hundred billion components, but that intelligence can be achieved by connecting a hundred billion ping pong balls!

Imitating the gourd, the gourd that is drawn can actually hold wine!

The promising quantum computer is therefore expected. It is said that once the quantum computer is successfully developed, the computing speed can be unimaginably improved. Professor Shi Yipeng of Peking University made an analogy. To determine if a cat is alive or dead, with dead being 0 and alive being 1, a traditional computer would need to calculate once for one cat, four times for two cats, and eight times for three cats, meaning the total number of calculations needed to determine the state of N cats is 2 to the power of N. However, a quantum computer would need only one calculation for any number of cats. The larger N is, the greater the advantage of quantum computers over traditional computers. It is said that a 50-bit (equivalent to $N = 50$) quantum computer can achieve a computing power of 1000 trillion times per second, far exceeding traditional computers. It is estimated that a quantum computer with a 1000-bit computing capability is enough to lift the veil on human consciousness. It is not a dream for artificial intelligence to have consciousness or will, but to make the dream come true, another dream, "quantum supremacy", needs to be realized.

Recent reports indicate that robots have been developed abroad which can perceive their own shape during work processes and automatically adjust their orientation for operational convenience. The ability to introspect and adjust oneself based on changes in the external environment is considered the budding of self-awareness in artificial intelligence. Technically, this might offer a pathway for AI development, though theoretically, it still needs to be demonstrated whether self-awareness is merely a simple perception of the relationship between the individual and the external environment, such as spatial distance.

According to the "Looking-Glass Self" theory in social psychology, human self indeed manifests as an awareness of one's relationship with the surrounding world, a process of establishing individual subjectivity. The self is gradually formed during childhood, requiring three coherent stages: seeing oneself from others' perspectives, imagining others' evaluations of oneself, and then forming corresponding feelings, such as pride or shame. This is considered the emergence of self-awareness. Clearly, the key here is not just perceiving the external world, but understanding it, and the ability to find meaning in the information coming from the external world.

Merely perceiving one's shape and its relationship with the environment and each other based on sensor-collected data and computations, without knowing the significance of such relationships, is still far from self-awareness or free will. If the threshold for self-awareness is lowered to this level, then there would be no animals in nature that do not possess "self-awareness".

1. Less optimistic

Scientists with this attitude are not so much pessimistic about the potential future development of artificial intelligence, as they are about the level AI research has already reached, "Machine will? Wishful thinking."

Michael Jordan, a renowned computer science and statistics expert, has clarified that many areas currently labeled as artificial intelligence are in reality machine learning, stating, "the AI revolution is far from here."

He believes that artificial intelligence aims to focus on "the advanced or cognitive abilities of human reasoning and thinking". However, after 60 years of development, high-level reasoning and thought remain elusive. "The development now referred to as AI mainly appears in engineering fields related to low-level pattern recognition and motion control, as well as in the field of statistics, which focuses on discovering patterns in data, making well-founded predictions, testing hypotheses, and making decisions".

In his view, what has made significant progress in the industrial and academic fields in the past 20 years is not AI, but IA, or "Intelligence Augmentation" (abbreviated as IA), which complements AI that imitates humans. Technologies from search engines, natural language translation, to audio and image generation are helping enhance human intelligence and creativity. These technologies are expected to encompass high-level reasoning and thinking in the future, but not yet.

Jordan even questions the notion that artificial intelligence simulates human intelligence, noting human intelligence is not perfect, with errors, biases, and other limitations, and lacks the ability to make large-scale decisions.

This view is essentially a direct undercut to the overly optimistic connectionist view: if the human intelligence being simulated has its limitations, what then can be expected of the simulated artificial intelligence?

As for those who argue that artificial intelligence may replace human thinking and creativity, he mocks, "If our goal is to build a chemical plant, should we first create a group of chemists and then have them figure out how to build it?".

In physical labor, humans have always acted according to the logic of "to do a good job, one must first sharpen one's tools". Will humans follow the same logic in the realm of intellect? Isn't developing software to design cars and then building robots to manufacture them exactly the pattern Jordan mocks?

If the extremely optimistic attitude borders on science fiction, the less optimistic attitude is too grounded. The history of technological development shows science fiction leading the way, with realism following. Looking up to envision the future and heads-down to push progress each have their place, matching and cooperating with each other.

2.10 Do Humans Need Artificial Intelligence to Have Free Will?

Do humans need machines to have will? From the perspective of purpose, perhaps not necessarily, but from the perspective of means, definitely.

Human society is complex, and choice is inevitable for anyone, even for individuals with intellectual disabilities. As artificial intelligence becomes more widely used and enters more areas of human life, fulfilling concurrent yet conflicting human needs, artificial intelligence will need to make choices. In meeting these diverse demands, forming a hierarchy of priorities or achieving a dynamic balance naturally becomes a human expectation.

When machine recommendations are criticized for being too vulgar, websites often excuse themselves saying "algorithms are neutral", but judges won't accept that. Algorithms need to navigate between personal preferences, corporate profits, and social benefits.

A unmanned drone embarking on a mission needs to balance two goals: completing the mission and preserving itself. If it can't complete the mission, preserving itself is meaningless; if it doesn't preserve itself, the mission will fail. Artificial intelligence must decide when to prioritize the mission, even at the cost of itself, and when to ensure its preservation first to attempt the task later. If it's a team operation, then the demand for making choices is even higher, requiring coordination and mutual support among multiple objectives, showcasing intelligence.

Human moral standards not only involve discerning right from wrong but also knowing how to choose when it's impossible to fulfill multiple valuable goals simultaneously. As artificial intelligence technologically simulates humans, it must emulate human wisdom in ethical dilemmas.

Moreover, when military robots are granted the autonomy to attack, it's essential to embed decision-making intelligence to ensure machines understand how to balance between eliminating enemies and protecting civilians, preventing autonomous attacks, malfunctions, or misuse.

As long as humans wish for artificial intelligence to be a creative executor, incorporating choice mechanisms into the algorithm's objective functions is an unavoidable step in AI's growth.

2.11 Can Artificial Intelligence Be Treated as Human?

Once machines have the capacity to make choices beyond direct human control, should they be held responsible if unacceptable consequences occur? How can machines assume responsibility?

It is no longer possible to view machines entirely as objects of rights, with all responsibility falling on humans, because the choices are made by machines, not humans.

It is also impossible to attribute everything to machines, as machines have not yet been recognized as subjects of rights, unable to bear responsibilities themselves.

At this point, does the law offer a "third path" allowing non-human machines to bear responsibility like humans? This requires the use of legal fiction as a legislative technique.

Article 16 of the General Principles of the Civil Law of the People's Republic of China stipulates: "In matters involving the protection of the interests of a fetus in inheritance, acceptance of gifts, etc., the fetus is regarded as having civil rights capacity. However, if the fetus is stillborn, its civil rights capacity never existed."

This provision is in clear conflict with the aforementioned Article 13. Article 13 stipulates that a person can only obtain legal subject status after birth, while Article 16 stipulates that an unborn fetus has already obtained legal subject status. This is not a conflict between laws, but involves "legal fiction".

Simply put, legal fiction is about treating non-A as A, treating "not" as "is". It was originally a legislative technique. According to Article 13 of the General Principles of the Civil Law of the People's Republic of China, a fetus is not a civil subject and does not have civil subject status because it has not been born. However, to protect the interests of the fetus, it needs to be treated as a subject in some cases. The most typical example of granting legal personality using legal fiction is the corporate system, with companies being the archetype.

Natural persons form an organizational body to conduct activities such as business or public welfare, which an individual cannot complete. Once officially recognized by the state according to legal procedures, it becomes a company, with an independent personality, and individuals are merely shareholders. A company is a legal person created by legal fiction.

Artificial Intelligence is about making computers perform various tasks that human intellect can do, and accordingly, the free will of artificial intelligence can only be discussed in terms of behavior. The mainstream view in the legal field is that robots do not possess the free will inherent in natural persons, nor do they have the capacity for will. Therefore, discussing the legal subject qualification of robots or artificial intelligence can only be done through legislative techniques, mainly by granting it legal qualifications through the method of legal "fiction".

The "fiction" technique, while granting artificial intelligence legal status, also raises a new issue: the implementation of legal responsibility for robots.

To what extent does the "fiction" of machines as humans facilitate the assignment of responsibility to machines, even if it is only temporary? Put plainly, what penalties should be applied to "criminal" robots: property penalties, deprivation of liberty, or the death penalty? Redemption through labor, forced upgrades, or direct destruction? Apart from affecting the machine itself, what other effects do any penalties have? To serve as a warning to humans or to deter machines?

While "fiction" may solve the legal obstacles of machine will, as long as machines do not truly possess free will, the legal means used to regulate humans will still face the dilemma of "heroes having no place to use their skills".

2.12 Will Human Will and Machine Will Wax and Wane?

Throughout history, human living conditions have been characterized by: houses for dwelling and vehicles for driving. In this relationship, humans have always dominated tools, using them for their purposes. Thus, the relationship between humans and machines is unidirectional, where tools may create dependence in humans but cannot dominate them. However, artificial intelligence is different from general tools. Within a certain range, artificial intelligence possesses its own independent capacity for action and can continuously improve or evolve through learning. There's reason to believe that this capability itself will continue to strengthen.

The autonomous vehicle is a technology platform, with its core technology being artificial intelligence, or the autonomous driving system. A key component of artificial intelligence is its adaptability to the environment. For instance, an autonomous vehicle limited to traveling between point A and B can perform significantly better on this specific route than on uncertain roads. This is because the vehicle, operating continuously on this route, becomes intimately familiar with the variations in road conditions during this period. As it accumulates more data, its performance becomes increasingly intelligent. Eventually, no human control is needed. One can simply get in the car and fall asleep, as the scenery has become mundane and aesthetically fatiguing. The machine becomes independent. If one day a human wishes to seek thrills against the machine's suggestions, the only necessity left is to find a new route in pursuit of fresh scenery. Purely in terms of travel efficiency, there is no point in humans driving themselves.

Some advanced forms of weak AI can even replicate parts of human intelligence, thereby changing the traditional unidirectional communication model between subject and object into a bidirectional one. Once machines develop emotional intelligence, such scenarios will become quite common. It can be expected that by this time it will be the turn of people to change. People will involuntarily regard objects that can communicate with them as their own kind. Not to mention artificial intelligence, even a dog, if you communicate with it a lot, you will use everyday language and even treat it as a family member. This is a reflection of human kindred thinking and empathy. Recognizing artificial intelligence as one of their own is a psychological shift in the face of artificial intelligence.

If psychology represents the unconscious state of the brain, then when the reality that "machines understand humans better than humans do" comes to pass, machines, by reading brain activity, could know what a person is thinking before the person is aware, and act accordingly. In such cases, if machine actions lead to unexpected outcomes, is it reasonable to hold humans accountable for the machines' overzealous actions? If the recorded brain activity indeed reflects such needs or motives, yet the person insists they were unaware of their own thoughts and had not explicitly instructed the machine to act, is it still reasonable to absolve them of responsibility?

So far, these thoughts remain at the philosophical level, and their value is nothing more than to spark people's imagination about machine will. Because the intelligent technologies that could potentially dominate humans are human inventions, machines dominating humans is merely an illusion of some people using machines to dominate others.

The issue is that as more people become dominated by machines, as algorithms gain a deeper understanding of individual preferences, as artificial intelligence becomes better at learning and autonomously evolving, and as those who wish to dominate others become increasingly trusting and dependent on machines, the following situation will inevitably occur: fewer people will be controlling machines, to a lesser degree, while more people will be dominated by machines, to a greater extent.

The battle of free will between humans and machines may not necessarily end with machines possessing free will. It could also lead to the weakening of human free will. Eventually, a critical point is reached where, relative to the weakening of human free will, even if machines do not have free will, the gray area between the two has expanded infinitely.

Physically, humans have already been proven to be weakened by machines, especially transportation devices. Competitive sports and fitness are human resistances to this trend. Intellectually, will humans be proven to be weakened by the invention of artificial intelligence?

All of these are possibilities, but what is more likely is that everything comes down to a matter of conceptual definition: if scientists could propose an operational definition similar to the Turing Test, could machine consciousness or will, like machine intelligence, become a recognized, accepted, and even recreated reality?

Chapter 3
How to Implement AI Responsibility?

On March 10, 2019, an Ethiopian Airlines Boeing 737MAX8 passenger plane crashed shortly after takeoff, with none of the 157 people on board surviving. This was the second severe accident involving a Boeing 737MAX8 aircraft in less than half a year, following the Indonesia Lion Air crash on October 29 of the previous year.

In the preliminary investigation following the Lion Air tragedy, investigators extracted data from the flight data recorder, commonly known as the "black box", revealing a life-or-death struggle between human and machine. In the final 11 min of flight, the plane's nose was pushed down more than 20 times under the command of intelligent software, with pilots repeatedly pulling it up until the plane ultimately lost control and crashed into the sea at a speed of 720 km per hour, killing all 189 people onboard.

This was not a robot rebellion in a science fiction movie, causing machine destruction and human death. Even a company as powerful as Boeing hadn't developed its intelligent software to such a science-fiction level. Due to structural reasons, the Boeing 737 MAX 8 aircraft is prone to stalling due to excessively high nose angles, endangering flight safety. Hence, an intelligent system was installed to adjust the horizontal stabilizer based on data from flight sensors if an improper flight attitude was detected, restoring the aircraft to a normal stance. This meant that there were two driving systems on the aircraft, manual and automatic, implying that both humans and machines had control over the same aircraft.

For safety, the aircraft was equipped with an intelligent system and procedures for pilots to turn off the autopilot and revert to manual control, which were included in the user manual. However, as long as the intelligent system wasn't turned off, any human effort to control the flight attitude would be counteracted by the system. In practice, this meant that as long as both operations coexisted, human strength was no match for the machine's capabilities. Regardless of what engineers intended, this arrangement objectively demonstrated that in critical flight attitude decisions, humans had ceded some control. In situations where both controls were active, machines making

J. Gu and C. Xu, *Will and Responsibility*, Human Intelligence,
https://doi.org/10.1007/978-981-97-3929-5_3

decisions for humans had become a reality. The result was an 11-min confrontation between human and machine, with the flight attitude changing repeatedly before tragically plunging to the ground.

As for why the failsafe switch from intelligent to manual operation did not work in both crashes, there have been various explanations. Official reports suggest that one of the two sensors was hit by a bird, leading to distorted data. Media reports have pointed to Boeing's lapses in training pilots for the new model, among other reasons. Before a convincing investigation conclusion is reached, speculation and taking sides are not scientifically meaningful.

What is unsettling is the strong symbolic significance of these two disasters, indicating that the scenario of humans and machines competing for control is increasingly close to us. Regardless of whether people are willing to accept it, even without discussing how to allocate power between humans and machines, how to implement responsibility for accidents caused by artificial intelligence has become an inescapable legal issue. Legislation on artificial intelligence is imperative, and to effectively implement machine accountability through legislation, we must return to the fundamental question of how we view artificial intelligence.

3.1 Where to Start in Implementing AI Responsibility?

Determining the responsibility of artificial intelligence necessarily rests on the fundamental judgment of this technology, with differing perspectives arising from different positions. Currently, there are two basic positions and three main viewpoints.

3.1.1 Tool Positioning and Dependency Theory

This viewpoint sees machines with artificial intelligence simply as tools of humans, which, regardless of technological advancements, are fundamentally dependent on and controlled by humans. Therefore, the ultimate goal of legislating robots is to harmonize the interests between humans, and discussions about machine responsibility, at most, have only nominal value. Essentially, it involves re-evaluating the responsibilities that different parties should bear in scenarios involving machines. In short, the issue of machine responsibility is only related to humans, not to machines themselves. Once human responsibilities are clearly allocated, the problem of machine responsibility is resolved.

In the main theme of the artificial intelligence tool theory, two variations can be clearly heard, optimistic and pessimistic.

The optimistic attitude stems from a simple belief. The history of humans is a process of continuously inventing tools, and technology can also be categorized broadly as a tool. The birth of each significant tool has always posed a huge challenge to humans, sometimes even giving the impression that the sky is falling. Practice has shown that the benefits outweigh the disadvantages, with the final outcome always being that humans overcome challenges and achieve greater progress. To some, this has almost become an "objective law"!

Artificial intelligence technology may bring greater challenges than before, which is undeniable. However, the notion that humans can dominate their own inventions and achieve self-improvement through technological advancement is equally unquestionable.

At the "2019 World Artificial Intelligence Conference" held in Shanghai on August 29, 2019, a conversation between Alibaba's Jack Ma and Tesla's Elon Musk was akin to a world-class "chicken talking with a duck". Jack Ma firmly positioned artificial intelligence as a tool and exhibited an equally firm optimistic attitude, whereas Elon Musk was not on the same track at all.

Jack Ma believes, "Many people are actually worried about artificial intelligence, but they need to have more confidence in themselves. Today we have many things without solutions, but there will be solutions in the future. We will find solutions, the younger generation will find solutions, and I am very optimistic". Therefore, Ma is optimistic, "I don't think artificial intelligence is a threat, I don't think artificial intelligence is something to be terrified of, because humans are smart, they will learn. We think it's good, it's interesting, we will embrace it".

Because of his optimism, Jack Ma is willing to call artificial intelligence "Alibaba Intelligence". Indeed, according to the acronym rule, both names can be abbreviated as AI, but whether one can substitute the other depends on the new meanings infused into the new interpretation.

The old company British Petroleum, abbreviated as BP, later used the abbreviation BP as its official name to reflect the company's values, and interpreted it as "Beyond Petroleum", meaning that the company contributes not only oil but also environmental protection to humanity.

Jack Ma's optimism is undeniable, but he does not provide a convincing argument for his optimism, merely stating, "99.99% of predictions made by humans in the past were wrong". Whether this generalization has statistical basis remains questionable. Even if it is true, does it also apply to overly optimistic predictions about artificial intelligence technology? If every generation pins its hopes on the next one being smarter and capable of solving the problems created by the current generation, thereby justifying their own unrestrained actions, it will inevitably be seen by future generations as a form of irresponsibility akin to the attitude of "after me, the flood!" For years, environmentalists have been calling for a solution to the excessive consumption of packaging materials and environmental pollution caused by the express delivery industry, but there has been no substantial improvement, and it may have been handed over to the next generation at China's largest e-commerce platform.

Even among those who also view artificial intelligence merely as a tool, many are deeply concerned about the misuse of existing technology today. Broadly speaking, this concern is that humans have developed technologies that should not have been developed, which could ultimately pose huge risks to humans. Just as the discovery of nuclear energy can be used for electricity generation to benefit humans, it can also be used to manufacture atomic bombs, placing the "Sword of Damocles" over humans' heads ever since. This has led to calls for banning the use of intelligent machines for combat purposes. It is absurd that humans who cannot control war can control the means of war.

On a smaller scale, pessimists worry that although artificial intelligence technology itself is neutral, once it is used by bad people, it will harm innocent people. Real-life examples include data leakage infringing on privacy, machine push to create information cocoons, "face swapping technology" being used for pornographic videos, and the "scissors hand" fingerprint being cracked and used for theft after taking a photo. Therefore, effectively preventing artificial intelligence technology from being misused has become a major concern of relevant legislation.

However, no matter what, the underlying message of these concerns is that artificial intelligence technology is not the problem. The people who develop or use the technology are the problem, because artificial intelligence is just a tool, and humans are the decisive factor. Legally, artificial intelligence can only be regarded as the object of rights, which means it does not have rights or obligations, nor does it bear machine responsibility.

3.1.2 Subject Positioning and Independence Theory

Independence theory doesn't claim that the goal of scientists is to develop artificial intelligence technologies with agency to rival humans. Instead, it suggests that regardless of human intentions, AI technology has its inherent logic and will inevitably evolve to produce robots with independent status. Concepts like "superintelligent AI" and "singularity" are based on this perspective and argument.

Another possibility is that the advanced development of AI technology may lead to a loss of human dominion in the relationship between humans and machines due to a decline in human capabilities, guiding humanity toward unintended outcomes based on machine logic.

In this regard, the rapidly developing algorithm push information in recent years has increasingly revealed the tenacity of machine logic. Providing information to individuals based on algorithms is known in the industry as "feeding". When machines cater to netizens' preferences and push information in a targeted manner, they show a strong tendency to be downward compatible with content, leading to increasingly vulgar information. The decisive role is the logic of maximizing click-through and conversion rates. As long as the internet continues to pursue traffic scale effect, seemingly neutral algorithms naturally have the effect of stimulating the beast side of human nature, which is "half angel, half beast".

From the standpoint of human-centeredness, to avoid imbalances in human–machine relationships and imbalances between humans caused by machines, two considerations are paramount. First, it's essential to prevent the creation of independent AI. Second, if independent AI is developed, adequate technical and legal measures must be prepared to regulate machine behavior and adjust human–machine relationships. Given AI technology's uniqueness, the legal measures involved may go beyond human legislative and judicial experience. Therefore, it is not only reasonable but also necessary to boldly imagine what challenges intelligent machines as a kind of existence with relative independence from humans will pose to the law, and to think outside the existing legal paradigms about the responsibility of intelligent machines.

3.1.3 Human–Machine Combination and "Human-Like" Theory

Between the simplistic view of artificial intelligence as a traditional tool and the radical view of artificial intelligence as an independent entity lies a middle ground. Simply put, artificial intelligence represents a new entity that can form an intrinsic connection with humans, necessitating a fresh perspective. Given the level of technology achieved today, artificial intelligence is neither a simple tool nor an independent entity. Artificial intelligence doesn't only operate on human-provided ideas but also generates its conclusions through perception and judgment. Essentially, scientists provide data, and artificial intelligence, utilizing algorithms, draws conclusions and decides on actions. It's challenging to classify it as either an independent entity or a traditional tool. If it is a traditional tool, then perception, judgment, and decision-making should come from humans. If it is an independent entity, these should be completed by the machine itself. Currently, humans contribute a part, and artificial intelligence forms a part, and neither can function without the other. This "grey existence" between two positions is the real value in discussing legal issues concerning robots, and naturally, it's also where troubles lie.

In fact, current artificial intelligence technology is rapidly advancing in the direction of human–machine coupling. Elon Musk, who has always been worried about the uncontrollable harm to humans caused by artificial intelligence technology, has announced the development of a technology that uses "sewing" to directly connect wires to the human brain, allowing computers to read information from brain waves. As this continues, humans and machines will merge in the thought process, making it difficult to determine whether future human actions are guided by free will or independently decided or "tempted" by machine intelligence, posing judicial challenges in assigning responsibility.

This viewpoint does not equate robots directly with humans, or suggests that robots should enjoy the status of citizen Sophia, and it certainly does not intend to replay similar "legal dramas" on the theoretical level. It views intelligent machines as "human-like"; not human, yet not traditional machines or "fabricated" humans by legislative techniques, but a "human-like" entity that lies between humans and traditional machines, which has not yet been precisely understood by humans.

3.2 Where Do Different Viewpoints Come From?

The stance one takes on AI technology—be it dependency theory, independence theory, or a middle-ground perspective—is influenced by the discussant's position, the current state of AI development, and existing legal thinking.

From a stance perspective, those holding a dependency view are typically technologists who have boundless faith in human capabilities. Based on their anthropocentric stance, they never accept the possibility of machines being parallel to humans, let alone surpassing them. The logic is simple: all technologies stem from experts' inventions, and whether from a theoretical, technological, or production standpoint, they haven't encountered any willful machines. Thus, ideas about machines becoming independent of humans are mere fantasy, and concerns over machines dominating humans are modern versions of unfounded worries.

Such optimism, while conducive to the development of AI technology, is evidently lax from a standpoint of thinking methods. Predicting the future based on human experience has always been a very risky thing. Assuming that what has happened before, and more than once, will happen again is often relegated to empiricism or commonplace thinking. New phenomena tend to defy empiricism, with genetic mutations occurring in ways that existing biological theories cannot explain, or else the influenza virus wouldn't catch medical experts off guard.

In the tech community, those who subscribe to the independence theory are generally experts with a humanistic spirit, highly sensitive and concerned about the human effects of technological advancements. They are optimistic about the technological prospects of artificial intelligence but are uneasy about the emerging shadows within those prospects.

Indeed, it's premature to grant intelligent machines a status of subjectivity based on current intelligence levels. However, the rapid progress of AI technology—not its widespread application success but the speed of underlying principle or logic advancements—necessitates forward-thinking about AI technology. Humans have never encountered a species "smarter" than themselves, and now not only have they seen it, but it is also created by humans themselves. Whether it can or will become "smarter" to the point of losing control, the initiative is still in the hands of humans. At the crossroads of technological development, whether to let technology run its course or apply an ABS-like approach, namely "point brake" car technique, to maintain the pace of research, giving artificial intelligence the necessary regulatory space without unleashing a genie from the bottle, warrants careful consideration. In this sense, rejecting citizen Sophia and proposing the concept of "human-like" is prudent and appropriate. Maintaining optimism for artificial intelligence technology as a tool while being cautious about intelligent robots as a purpose is not contradictory but complementary, ensuring the healthy development of artificial intelligence technology at a deeper level.

Legal discussions about AI legislation predominantly start from the tool positioning, not because of individual opinions but due to the discipline itself. Legal people tend to speak within the bounds of existing legal provisions, avoiding statements without legal basis, as it reflects their professional quality. After all, for practitioners, it is difficult to win a lawsuit if they walk in front of the law. Since the law does not provide a right subject status for artificial intelligence, legal professionals naturally adhere to principle of "machine equals tool".

To put it more thoroughly, many legal people also adopt a tool-like attitude towards the law itself, viewing it merely as a means to regulate human-to-human interest relationships. Humans have interests, but machines inherently do not; they merely reflect human interests. Therefore, the discussions about machines are only appropriate when conducted within the framework of human interests.

This perspective itself is not problematic, and in the context of the law lagging significantly behind the development of artificial intelligence technology, it can also contribute to the advancement of relevant legislative construction and is worth exploring in depth. However, its clear shortfall is equating artificial intelligence technology or intelligent robots with traditional tools, failing to focus on the unique characteristics of intelligent technology and its profound implications for human existence. This leaves a certain degree of internal space in legal thinking underutilized, making it difficult to provide foresight for the further development of artificial intelligence technology and to propose breakthrough directions for legal construction in the era of artificial intelligence.

Relatively speaking, experts who view the law as an end in itself and excel in theoretical contemplation are more inclined to adopt an independent or "anthropomorphic" stance towards artificial intelligence. In addition to the public or personal issues that artificial intelligence technology may bring, the problems that the law itself may face due to the development of artificial intelligence technology receive more of their attention. Even considering the interest factors in the legal field, for them, it is the overall interest of mankind and public welfare, not the commercial interests of individual manufacturers, users, or service providers of intelligent machines.

On the whole, the tendency towards instrumentalism or legalism is quite common among domestic legal practitioners, who are more skilled and accustomed to viewing AI technology from a legal perspective than viewing the law from the standpoint of intelligent machines. Faced with issues that diverge from existing legal thinking, there's a natural impulse to force the proverbial robot's foot into the current legal boot, which not only overlooks the legal effects that AI technology might bring but also wastes opportunities for legal construction, including breakthroughs in thinking.

The issue now is that while theoretical discussions can be lengthy, with "truth emerging from debate", practice waits for no one. As artificial intelligence technology develops and legal issues in application scenarios continue to arise, there's an urgent need for immediate legal solutions, as the judiciary must provide clear rulings on AI-related civil cases regardless of the presence or absence of laws. This urgency seems particularly pronounced in the nascent field of autonomous vehicles, potentially serving as a window into the legal effects of artificial intelligence.

3.3 What Legal Issues Do Autonomous Vehicles Raise?

The hottest topic in the field of artificial intelligence today is undoubtedly autonomous driving, which is due to both a demand in everyday life and a necessity from the technology itself: whether it's a mule or a horse, take it out for a walk, only then can the public believe that artificial intelligence technology really works. A successful autonomous vehicle is far superior to the most powerful AlphaGo.

Autonomous vehicles are composed of three units: perception unit, decision-making unit, and control unit. The perception unit is the sensor and perception algorithm. The perception unit includes sensors and perception algorithms. The decision-making unit encompasses the mechanical control circuits, controlling machinery, and data analysis. The control unit, for example, the control interface, brakes, and steering wheel, can make decisions based on this information. These are the three components of autonomous driving. Self-perception, independent judgment, and autonomous action distinguish autonomous driving from traditional robots, and when accidents occur, the legal issues involved are related to these aspects.

3.3.1 The World's First Fatal Autonomous Driving Case

There was an American who was an "autonomous vehicle enthusiast", extremely fond of autonomous cars. As soon as the autonomous Tesla Model S came out, he bought it and drove it on the street. Unexpectedly, he set a world record—the world's first fatal accident involving an autonomous car.

One day in May 2016, he was sitting in the driver's seat of the autonomous car, not putting his hands on the steering wheel as usual. If it's autonomous driving, why would it need human control? The sun in Florida was quite strong, and a white truck was coming in the opposite direction. The car's perception system failed to recognize the incoming truck, and the car, in "autonomous mode", didn't swerve or brake but went under the truck, killing the man inside instantly.

Then came the lawsuit, landing on the judge's desk: in the event of a fatal accident involving an autonomous vehicle, who is responsible?

If the car owner is blamed, there's an issue. The owner trusted in autonomous technology, believing that there was no need to hold the steering wheel in autonomous mode. Isn't it redundant if autonomous driving still requires manual control?

If the liability falls on the company or producer of the autonomous vehicle, that too is problematic. The manufacturer might argue, "Although it's called an autonomous vehicle, it hasn't achieved full autonomy yet, and the driver still has a duty to keep hands on the wheel during operation. You don't have to control the direction constantly, but should be ready to take over the vehicle at any time. After the autonomous system issues a warning, the vehicle must be manually controlled within 6–12 s."

If the liability is attributed to the other vehicle's driver, there's also a problem. The truck didn't violate any traffic laws. It was the autonomous car that crashed under the truck, not the other way around.

Who should bear the responsibility in the end?

After a 6-month investigation, the National Highway Traffic Safety Administration (NHTSA) in the United States finally determined: Tesla's Autopilot system had no defects. All necessary warnings were given, and all required operations were available; there were no defects. However, by using the term "autonomous driving", Tesla exaggerated the car's capabilities, misleading drivers to trust the vehicle beyond its limits.

Though the NHTSA made two judgments on this accident, it still didn't clarify who should bear responsibility.

In response, Tesla's boss Elon Musk said, "Our current autopilot has been proven to provide better safety than manual driving, with a road death rate that is one-third of the world average. Even if our system is only 1% better than manual driving, we could save 12,000 people out of the 1.2 million who die in car accidents each year. So I think it's morally wrong to shelve a technology that can improve safety just because of fear of public opinion or legal prosecution".

Musk only emphasized the societal value of autonomous driving, suggesting that the benefits it brings to individuals and society at large far outweigh manual driving, hence, rolling out autonomous driving is meaningful. This is the thinking of a technology innovation entrepreneur.

However, the law does not consider issues this way. While purpose matters, the law's role is to set rules to adjust interpersonal relationships, striking a balance between technological innovation and human safety. Both "fish and bear's paw" are needed, the key is how to adjust the law to effectively balance technological innovation and social benefits.

3.4 How Does Law Deal with Autonomous Car Accidents?

In the face of the liability issue of autonomous vehicles brought about by the Tesla accident, within the current legal framework, there are roughly two paths of liability: motor vehicle damage liability and product liability. Let's first discuss the pros and cons of the motor vehicle damage liability path.

According to existing laws, responsibility for the accident must be investigated based on the relationship of rights and obligations. Legally, an obligation refers to what an individual should do. Obligations are relative to rights. With certain rights come certain obligations, and only when these obligations are violated does responsibility arise.

For example, if a consumer makes an online purchase, paying 1000 yuan, but doesn't receive the ordered goods, there's a problem. Under the contract, the consumer has fulfilled his obligation by paying 1000 yuan. If the seller fails to deliver the goods, they violate their obligation, thus entering the realm of responsibility. Violation of obligations necessitates bearing responsibility.

In civil law, the most important principle related to responsibility is that people are only responsible for their own actions. If a seller fails to deliver, the consumer can't hold the seller's father accountable, expecting the elder to resolve it. The father would likely respond, "This is what my son promised you, not what I promised you. He is over 18 years old and an adult. He is responsible for himself". This illustrates the legal principle that individuals are responsible only for their own actions, not for anyone else's, except those under their guardianship. Further, the law stipulates that individuals are responsible only for their wrongful actions; there is no responsibility without fault. For instance, if someone is walking a dog on a leash and a passerby, finding the dog adorable, decides to tease it but gets bitten, the dog's owner would typically be held responsible. However, in this case, the passerby might bear fault for provoking the dog, potentially absolving the owner of responsibility. Generally, individuals are responsible for their faults. Yet, sometimes responsibility must be assumed even without fault.

When driving a vehicle on the road and a collision with a pedestrian occurs, the driver is responsible regardless of fault due to the "steel body" and speed of the vehicle. Even if all actions comply with traffic laws, the driver bears responsibility, known as "no-fault liability".

According to the current "Road Traffic Safety Law of the People's Republic of China", the legal responsibility of traffic accidents is usually handled in this way.

Speaking of the law, there is a familiar saying: based on facts and guided by law. Litigation must refer to the law, and when dealing with road traffic accident liability, the most important provision is Article 76 of the "Road Traffic Safety Law of the People's Republic of China": "If a motor vehicle causes personal injury, death, or property loss in a traffic accident, the insurance company shall compensate within the limits of the compulsory third-party liability insurance for motor vehicles; the insufficient part shall be borne according to the following provisions: (1) In the event of a traffic accident between motor vehicles, the party at fault shall bear the compensation liability; if both parties are at fault, they shall share the responsibility according to their respective faults. (2) In the event of a traffic accident between a motor vehicle and a non-motor vehicle driver or pedestrian, if the non-motor vehicle driver or pedestrian is not at fault, the motor vehicle party shall bear the compensation liability; if there is evidence to prove that the non-motor vehicle driver or pedestrian is at fault, the compensation liability of the motor vehicle party shall be appropriately reduced according to the degree of fault; if the motor vehicle party is not at fault, they shall bear no more than 10% of the compensation liability".

In daily life, this provision is basically applicable to every case. According to this provision, when an accident causes injury, death, or property loss, regardless of fault, the insurance company initially covers the damages. The remaining liability is then determined by both parties based on the circumstances.

Specifically, liability is allocated in two scenarios. First, if it's a collision between vehicles, it depends on which vehicle is at fault. If both are at fault, the extent of each vehicle's fault determines the share of the damages. Second, in collisions between a vehicle and a pedestrian, the vehicle is responsible regardless of fault. If the pedestrian is at fault, their fault may reduce the vehicle's liability. In cases of vehicle–pedestrian collisions, the driver assumes responsibility, irrespective of fault.

3.5 What Adjustments Need to Be Made in Legislative Technology in the Face of Autonomous Driving?

Once artificial intelligence technology is adopted and an autonomous driving system is introduced, the situation changes. There is an important conversion here. The manufacturer of this system is also the driver. Now autonomous vehicles are not as advanced as imagined. They cannot operate entirely without a human, necessitating a driving seat and steering wheel. There must be someone seated in the driver's position, with the obligation to keep their hands on the wheel. Since under normal circumstances it's not the person but the artificial intelligence system driving, the real driver is not the person seated at the driving position but the manufacturer of the intelligent system. After this conversion, the producer becomes the driver. According to Article 76 of the "Road Traffic Safety Law of the People's Republic of China", it is the producer's turn to bear responsibility.

The producer becoming the driver sounds logical, but in legal terms, being reasonable isn't enough. There must be a legal basis. Legal, reasonable, and compassionate—legality comes first. Let's look to the "German Road Traffic Act" for a precedent, such as Article 2, Section 15, Sentence 2, which legally defines driving instructors as drivers in the context of driving tests. If an accident occurs while a driving instructor accompanies a learner during a test, the instructor should be regarded as the driver. In fact, for a person learning to drive to reach the level of a qualified driver, the instructor must bear a certain responsibility. German law provides a reference where instructors, though not drivers by default, can be legally considered as such. By analogy, viewing the producer as the driver finds its rationale. Legally, this type of technical handling is called "fiction".

The essence of this solution is to regard the producer of the intelligent system as the driver, to hold them accountable for road traffic accidents. Following this strategy, most issues currently posed by autonomous driving could be resolved. However, this also leads to another question.

3.6 What Obligations Do Natural Persons Have Under Autonomous Driving Mode?

The aforementioned solution, by "fabricating" the intelligent system as the driver, in plain terms, treating the machine as a human, successfully shifts the responsibility from the driver to the manufacturer of the autonomous vehicle. Although such a transformation can stand legally, it is still difficult to get manufacturers, especially creative individuals like Elon Musk, to accept it. With the current level of technology, do autonomous vehicles really not require human involvement? Should the driver not be held accountable when the intelligent system malfunctions and causes an accident? Can a natural person completely offload their responsibility onto a machine, and can a natural person really use the concept of a "fabricated" person to escape accountability?

Let's examine the current requirements for drivers of autonomous vehicles.

Table 3.1 fully lists all levels of autonomous vehicles, especially detailing the corresponding obligations that drivers, as natural persons, should fulfill at different levels of autonomous driving.

L0 (Manual Driving): The operation is conducted by a human driver, surrounding monitoring is by the human driver, and takeover is also by the human driver. This means an individual driving a car must fulfill all duties and react promptly in emergencies. Since it's human-driven, the machine has no role here.

L1 (Assisted Driving): Automatic cruising. Also known as cruise control, where after a certain time post-activation, the vehicle can travel at a set speed even after the driver releases the accelerator, until the driver accelerates or brakes. L1 represents a level of assisted driving functionality.

L2 (Partial Automation): The vehicle is driven by the vehicle itself, while the surrounding monitoring and takeover are done by a human driver. During the journey, humans need to supervise and watch for any emergencies. If any occur, they need to take over immediately. Therefore, the driver's hands must remain on the steering wheel, just like when driving manually.

At the L3 (Conditional Automation) stage, the vehicle operates the drive and monitors the surroundings, but the human driver takes over. Here, the human is somewhat removed from driving functions. The intelligent system operates and monitors the drive. Human obligations are significantly reduced but still present. The driver only needs to keep their hands on the wheel and take over immediately upon receiving an alert from the intelligent system.

The current autonomous driving intelligent systems are roughly between L3 (Conditional Automation) and L4 (High Automation). L4 (High Automation) and L5 (Full Automation) are truly autonomous driving, where everything is managed without human intervention. By then, drivers bear no responsibility, and all liabilities fall on the manufacturer, as both harm liability and product liability are inescapable for the producer.

3.7 How to Attribute Responsibility Along the Product Liability Path?

In daily life, product liability can be understood as the user having no responsibility, with the root of the problem lying in the product itself, not improper use.

Table 3.1 Autonomous driving levels

Autonomous driving classification		Name	Definition	Operation	Surrounding monitoring	Takeover	Application scenarios
NHTSA	SAE						
L0	L0	Manual driving	The human driver is responsible for all driving tasks	Human driver	Human driver	Human driver	None
L1	L1	Assisted driving	The vehicle provides driving assistance for either steering or acceleration/ deceleration; the human driver is responsible for all other driving tasks	Human driver and vehicle	Human driver	Human driver	Limited scenarios
L2	L2	Partial automation	The vehicle provides driving assistance for multiple operations related to steering and acceleration/ deceleration; the human driver is responsible for all other driving tasks	Vehicle	Human driver	Human driver	
L3	L3	Conditional automation	The vehicle carries out the majority of driving operations, while the human driver must remain attentive to take over if necessary	Vehicle	Vehicle	Human driver	

(continued)

Table 3.1 (continued)

Autonomous driving classification		Name	Definition	Operation	Surrounding monitoring	Takeover	Application scenarios
NHTSA	SAE						
L4	L4	High automation	The vehicle performs all driving operations, while the human driver does not need to maintain attention. However, it is subject to specified road and environmental conditions	Vehicle	Vehicle	Vehicle	
	L5	Full automation	The vehicle performs all driving operations, while the human driver does not need to maintain attention	Vehicle	Vehicle	Vehicle	All scenarios

A consumer buys a TV, and while watching it, the TV explodes. Unfortunately, there are guests visiting who are also injured by the explosion. If the host did not operate improperly or do anything else, this injury event involves product liability, for which the producer is responsible.

For accidents involving autonomous vehicles, the cause can vary, including brake failure, steering failure, etc., with the most typical, significant, and challenging cause being a problem with the algorithm. Under the thinking path of road safety accidents, the producer is always responsible because the intelligent system acts as the driver. However, under the product liability path, this may not necessarily be the case. If the producer can prove that the product has no defects and the AI algorithm is without fault, they bear no responsibility.

It might be puzzling how the law can seem ambivalent. Choosing the first approach, the producer bears responsibility; choosing the second, they don't. Isn't that inconsistent?

Yes, in legal terms, this belongs to litigation strategy. If the driver chooses the road safety path, they may win, and the producer has to compensate; if they choose the other path, the driver has to pay a heavy price. Because if the producer can prove that their product and algorithm are fine, and the car has no defects, they bear no responsibility.

For autonomous vehicles, the key lies in how algorithm defects are determined. There are many defects in car products stipulated by law, such as manufacturing defects, design defects, and warning defects. This discussion focuses only on design defects. If the programmer leaves a defect when designing the algorithm, and it can be proven, then the producer can be held directly responsible. The problem is that algorithms have unknowability, even the designers find it difficult to explain, which is the so-called "algorithm black box". Not to mention that ordinary judges do not have the ability to intervene in the algorithm, even the programmers themselves do not have the ability to intervene in the "algorithm black box", and it is uncertain which layer of the algorithm has a problem. In the use of artificial intelligence technology, the biggest problem faced after a problem occurs is that the algorithm is not interpretable.

When autonomous driving causes harm and it cannot be determined whether there is a defect in the algorithm, the court still needs to make a responsibility judgment and require the producer to improve the design system based on this. The method is brutally simple: if the producer can't clarify, it indicates a defect. By reversing the burden of proof, if the producer cannot explain, the product is judged defective.

The law is so crude that it is difficult to achieve an ideal effect that is win–win for all parties. After all, besides closing a lawsuit, the law also needs to make a total social value measurement, especially considering the social value of technological innovation. If the approach is just about simplicity, the legal profession would surely be among the first to be replaced by artificial intelligence.

3.8 How Can Producers Protect Themselves Through Development Risks?

If a judge chooses to apply product liability when the producer of an autonomous vehicle cannot prove that the product's algorithm is flawless, it would be overly harsh for the producer, adversely affecting technological innovation and its application. Therefore, the law needs to continue to balance interests, offering producers a reason to be exempt from liability. This system is known as the "development risk defense".

On the one hand, if a product is defective and causes personal or property damage, the producer should be liable for compensation. On the other hand, if the producer can prove any of the following situations, they can be exempt from compensation responsibility: first, the product was not put into circulation, i.e., the product was not marketed but was used by someone; second, at the time the product was put into circulation, the defect causing the harm did not exist, meaning the harm occurred because the user mishandled the product; third, the level of scientific and technical knowledge at the time the product was put into circulation could not detect the defect's existence. This third condition is the development risk defense, implying that technology is always advancing, and producers cannot anticipate defects that have not yet been discovered.

Referring to this, one can identify a key feature of artificial intelligence. Autonomous vehicles must interact with their environment at all times, and their intelligence level gradually improves after being put into use, potentially exceeding the original expectations of the designers. This means that even the most diligent scientists and meticulous programmers cannot fully predict future actions of the artificial intelligence. Therefore, the proof reversal method for product defects cannot be simply applied.

However, if this is the case, whenever an autonomous vehicle causes harm, the producer can use this defense, making it seem as if the development risk defense was specifically designed for inexplicable algorithms. The law first imposes a harsh responsibility on the producer, then offers a broad defense. Does it want the producer to be responsible or not?

3.9 Can Recall Obligations Solve the Dilemma?

The developmental space should be left for technological innovation, while producers should not take chances and lack the initiative for progress, ultimately leading to significant social risks. The law needs to employ a clever strategy.

The law stipulates that if any product is found to have defects after being put into circulation, the producer and seller should take timely remedial measures such as warnings and recalls. If timely remedial actions are not taken or if the measures are inadequate, resulting in harm, they must bear tort liability. Once an accident occurs with an autonomous vehicle, or if the producer realizes the algorithm or other designs are not suitable for the current situation before an accident happens, they have a duty to recall. According to this regulation, producers of autonomous vehicles are obligated to update the technology in a timely manner when it becomes outdated, especially in updating and upgrading the algorithm, which is essentially a form of recall.

The two legal paths surrounding the liability for harm caused by autonomous vehicles are solutions necessitated by the current law, which has not yet created special rules for AI driving systems. It is hoped that when autonomous driving reaches levels L4 and L5, artificial intelligence will have entered the stage of strong artificial intelligence, and its application field will be broader. At that time, recognizing the subjectivity of artificial intelligence and letting autonomous vehicles take responsibility may become legally feasible.

There are many ways to make autonomous vehicles responsible, such as setting up insurance, guarantee funds, reserve funds, etc., to ensure that the responsibility is fully and effectively covered. This approach better serves ordinary consumers and encourages producers to update technology, while the responsibility ends with the autonomous vehicle and does not extend to producers, designers, sellers, and users. Of course, it is not difficult to see that all these legal resources are fundamentally based on humans, not really focusing on the future situation where machines possess intelligence and become independent entities.

3.10 Is Recall Itself Also a Dilemma?

Humans hope for technology without flaws, but it's impossible to develop technology without any. Some flaws are known yet unavoidable, and some cannot even be recalled.

Judging product defects requires standards, and the necessity of mandatory recalls must be supported by law. However, legal provisions regarding product defects must consider not only the producer's obligation towards consumers but also the producer's capacity to endure. No product in the world is without defects, and addressing these defects involves considerations of technological feasibility and financial balance.

In terms of car safety, one must consider the convenience cars bring. Safety cannot come at the expense of convenience. Otherwise, staying at home is definitely safer than sitting in a car, driving at a speed of 1 km per hour is definitely safer than 100 km per hour, and sitting in a tank is definitely safer than sitting in a car, regardless of fuel consumption. All of these fall outside the "reasonable expectations" of demand. Secondly, compared to the president's car, ordinary family cars are not up to standard in terms of safety. To achieve the same level of safety, an ordinary consumer would either have to spend much more money or become the president first.

The reason why Musk used the death rate of manual driving as a defense for autonomous vehicles killing people is that he used the concept of reasonable expectation: manual cars have a much higher death rate, and the public can still tolerate it, so why fuss over the occasional deaths caused by autonomous vehicles?

Therefore, how defects are legally defined and what triggers a compulsory recall need to be clearly established through specifically designed technical standards and identification methods. The law sets safety production standards while tolerating "under-safety" within the bounds of reasonable expectations. This means that a product, such as a car, even if it meets safety standards, is not necessarily without defects. The key is whether the defect exceeds the consumer's safety expectations for similar products.

There's a legal theory abroad that advocates for determining product defects through a "cost and benefit" calculation of alternative products. If the use of alternative products can bring better results without significantly increasing costs or significantly reducing benefits beyond the producer's capacity, then the alternative is desirable, the defect exists, and recall is necessary. Otherwise, the defect is not recognized, and recall is not mandatory.

China was late in implementing a recall system for the automotive industry. This led to situations where some foreign brand manufacturers recalled defective cars worldwide but refused to recall the same type of cars for Chinese consumers. This caused Chinese consumers to feel indignant, believing it to be open discrimination.

In fact, this situation has nothing to do with discrimination, but is due to the lack of a defect product recall system in China at that time.

The real reason for the delayed establishment of a recall system in China was not to facilitate foreign enterprises but to leave room for the development of domestic cars. At their inception, domestic cars were inevitably flawed. Immediate recalls for any issues would have doomed the producers. Thus, the risk associated with defects had to be partially shifted to domestic car buyers and users, with accidents being offset by insurance or societal funds.

There is not only a balance of interests but also a balance of time. Under the protection of laws and policies, domestic cars have developed, and their ability to bear costs has also strengthened. At this time, the recall system must be established in a timely manner, otherwise it will not only result in unnecessary loss of consumer interests, but also encourage a reliance on the system by domestic car manufacturers, hindering technological progress. Therefore, whether to recall or not is not only related to the interests of a certain group, but also requires overall coordination.

In fact, in the context of China's severe aging and even super-aging trend, and the gradual disappearance of the demographic dividend, intelligent robots that provide life services have a greater social demand than autonomous vehicles. More and more elderly people lack care from their children, making it impossible for a pair of working-age young people to take care of 4 or even more elderly people while raising their children. On the other hand, they can't enter elderly care facilities because there are simply too many people to accommodate. In this situation, if there are robots that can provide service functions, it is undoubtedly a great merit. However, with the current level of intelligence and mechanics, if a robot causes harm to the elderly due to defects during the service process, how should the responsibility be determined?

Blaming the elderly is unreasonable and inhumane. Blaming the machine is unimplementable. Blaming the children who bought the robot to provide services for the elderly may inevitably affect their purchasing enthusiasm and ultimately exacerbate the problem of elderly care. Blaming the producer would hinder the supply and improvement of service robots.

Given the imbalance between resource supply and social demand, it is essential to design legal systems that better balance different interests. The aim should be to allow technological advancements to benefit the elderly and ease the burden on their children, while ensuring human safety, particularly safeguarding the rights of the elderly as a vulnerable group, and maintaining the producers' motivation. Encouraging technological innovation and the integration of robots into homes and nursing facilities to undertake tasks that humans can do but lack the time for, allowing more elderly to enjoy their twilight years, and enabling young people to lead independent lives, represent the right direction.

Therefore, even if a home service robot infringes rights, the issue of responsibility attribution must take into account the degree of aging, the development of elderly care services, and the technical level reached by robots. Only after considering these factors can the extent of responsibility borne by the producer, the "unlucky consumer", and society at large be determined. In cases of robots causing harm to the elderly, the necessity for widespread recalls, at least initially, requires careful consideration.

3.11 Hold Machines Accountable or Exempt Individuals?

The discussion above, while seemingly reasonable and clarifying the distribution of responsibility in the application of AI driving, actually says nothing because the focus is still on humans, not machines. Legislation on AI responsibility, in the name of holding machines accountable, often ends up merely exempting producers, purchasers, or users from liability, without placing any real responsibility on the machines themselves.

This perspective on addressing machine responsibility stems from two sources: firstly, the current state of AI technology development, where machines are not yet capable of bearing responsibility. Any attempt to prematurely assign responsibility to machines, no matter how well-designed, is bound to fail and find no proper place for accountability. Secondly, given that machines are still created by humans, all responsibilities ultimately cannot detach from human involvement. As for how much responsibility humans should bear for machine actions and their negative outcomes, despite the numerous theoretical deductions possible, accountability in practice can only be sought from humans.

These two reasons lead to a situation where, at least for now, theoretical deductions can advance rapidly, almost to the point of science fiction, while practical progress must proceed step by step, and often seems to be standing still. What truly matters is the exploration and discussion on how to redefine and redistribute responsibilities that change in nature due to the widespread application of AI technology, at both the individual and collective human levels. Insisting on traditional views of liability when individual free will is encroached upon by artificial intelligence will not help solve the new problems posed by intelligent devices and might, in fact, hinder the development or control of AI technology.

Elon Musk's defense of the Tesla incident, while certainly aimed at absolving responsibility, aligns with the general logic of capital. However, the distinction and connection he makes between individual interests and social benefits in determining responsibility offers valuable insights for future AI legislation. "The person who tied the bell must untie it". Problems created by humans can only be solved through human improvement in the end.

In summary, the implementation of machine responsibility is a dynamic process that will mature and refine with the development of AI technology. The law need not be overly prescriptive, but forward-thinking is essential. It's crucial to recognize that solutions found within the current legal framework may not necessarily be the ones adopted in the future. Especially when strong AI is defined as equivalent to human intelligence, the effectiveness of legislation from a human-centric standpoint is questionable. Whether a dual legal system becomes a more viable option remains a significant area for discussion.

Chapter 4
How Do Robots Balance Rights and Obligations?

In everyday use, the same concept of responsibility often carries different meanings in different contexts, with both overlaps and significant differences between them. However, in theoretical discussions, experts generally agree on two main uses.

First, responsibility is categorized as a moral concept, where an individual with free will acts consciously or unconsciously, causing certain results, and thus must bear corresponding responsibilities. This responsibility can be demanded by the community based on existing consensus, or it can be self-imposed by the individual based on self-cultivation. The two are usually not contradictory. For example, when we praise someone, we might say, "This person is really responsible", which often implies affirmation of this person's adherence to social rules and self-discipline. This is a broad interpretation of responsibility.

Second, responsibility is viewed as a legal concept, where natural persons or legal entities must bear corresponding responsibilities due to failure to fulfill legally stipulated duties or faults in fulfilling those duties. Here, responsibility can be related to an individual's subjective intention, or it can be less directly related. The law recognizes "fault", conveying that improper behavior, even if unintentional, cannot be completely exempted from responsibility. Since responsibility is invariably associated with obligation, without obligation there is no responsibility, and obligation itself must be equivalent to rights. Therefore, responsibility is also related to rights.

Clearly, the previous discussion within the scope of free will focuses on moral responsibility, while the issue of responsibility allocation in the application scenarios of artificial intelligence mainly involves legal responsibility. For humans, moral responsibility offers greater freedom, while for robots, legal responsibility has more practical significance. Therefore, it is more necessary to clearly define the concept of legal responsibility, clarify the conceptual chain from responsibility to obligation, and then to rights, and the logical connections therein when considering robot responsibility.

J. Gu and C. Xu, *Will and Responsibility*, Human Intelligence,
https://doi.org/10.1007/978-981-97-3929-5_4

Since rights, as the exercise of free will, are still exclusive to humans and not shared by any non-human species or objects, this presents an unsolvable theoretical dilemma. As long as robots are not recognized as having free will or rights, it naturally follows that robots are not acknowledged to have obligations, making it impossible to implement legal responsibilities for robots. All attempts at AI legislation will then revolve around humans, struggling to make substantive breakthroughs and expansions.

While legal scholars grapple with the bottleneck of established thinking, science fiction writers have proposed their visions long before the explosive development of AI technology and even before the appearance of any intelligent machines. The "robot laws" or "robot principles" they've designed, although purely speculative about an "unborn" "new intelligent species", reveal the thought processes, limitations, and blind spots humans have when approaching these issues. Serious artificial intelligence legislation will not imitate, but will refer to them. "Man, know thyself." In artificial intelligence legislation, as in many other occasions, recognizing our own nature is wisdom that ensures our success.

In the modern society where technology is rapidly developing, extraordinary science fiction writers, with their imagination, are not only good friends of scientists, but also good friends of legislators. However, before inviting these pioneers to the stage, it is worth warming up with some basic legal knowledge for them.

4.1 How Does Law Define Responsibility?

In everyday language, the statement that someone "has a responsibility" to do something equates to having an obligation to do it, and ordinary Chinese people are not adept at clearly distinguishing between the legal and moral attributes of responsibility or obligation.

In fact, the broad sense of legal responsibility refers to the obligation of all organizations and individuals to abide by the law and consciously maintain the dignity of the law. In other words, it refers to the responsibility of natural persons or legal entities to the law itself.

The narrow sense of legal responsibility refers to the compulsory legal responsibility that violators should bear for their illegal acts. Legal responsibility is linked to illegal acts, meaning that natural persons or legal entities only need to assume corresponding legal responsibilities when they engage in certain illegal actions. For some illegal acts, regardless of whether they result in improper consequences, the nature of the responsibility does not change, but the extent of the liability pursued does. Drunk driving does not necessarily lead to traffic accidents, but to prevent potential dangers, the law stipulates that as long as someone drives under the influence of alcohol, and if a breath or blood test shows an alcohol content of more than 20 mg per 100 mL of blood, it is defined as drunk driving, and if it exceeds 80 mg, it is considered severe drunk driving. Exceeding these standards, the driver will be subject to penalties according to relevant laws, including points deduction, fines, temporary suspension of the driving license, detention, or even sentencing.

Legal responsibility not only has clear and specific regulations but is also enforced with the coercive power of the state, pursued by state-authorized agencies according to the law, and subject to legal sanctions as stipulated. No other organization or individual has the right to interfere with this power.

If someone is caught stealing on the street, passersby can intervene on the spot, and assist in capturing the thief if they attempt to flee, but they cannot arrest, beat, imprison, or interrogate the thief. Acts of bravery that exceed the legal limits are also legally accountable, and the penalties received may be more severe than the theft itself.

The importance of legislating for robots lies in defining responsibilities between humans and machines with the guarantee of state coercion, which can only be clarified through legislative procedures to determine the responsibilities of all parties involved, including robots. Then, according to the regulations, the state can pursue the liability of intelligent machines or exempt certain legal entities from responsibility.

Legal responsibility, according to the nature and degree of the behavior and its consequences of natural persons or legal entities, is specifically divided into criminal legal responsibility, civil legal responsibility, administrative legal responsibility, etc. In the current application scenarios, artificial intelligence is more likely to cause civil torts due to immature technology. Even if personal injury incidents occur, they are usually due to technical defects, not intentional, and with a limited impact. Therefore, the discussion about robot responsibility can be limited to the civil domain. As for the future threats that super artificial intelligence may pose to humans, it can be left for later discussion.

However, some experts argue the opposite, believing that the criminal nature of responsibilities has more value for discussion. After all, those who are apprehensive about the prospects of artificial intelligence technology are most afraid of the citizen Sophia's blunt and horrifying answer when asked "what do you most want to do": "Destroy humans!" Although merely a marketing stunt by Hanson Robotics, it reaffirms the fears of many experts already anxious about the rapid development of artificial intelligence.

Therefore, the legal responsibility for the most severe crimes possibly committed by robots—"anti-human behavior"—must be included in the discussion, even necessitating a start from this crucial aspect of robot responsibility with significant implications for humans.

4.2 Why Do Rights and Obligations Need to be Equivalent?

In legal terms, rights and obligations must be equivalent. There are no obligations without rights, and no rights without obligations.

The term "rights" refers to the power and interests enjoyed by citizens according to the law, or the power and interests autonomously enjoyed by the subjects of legal relations within the scope prescribed by law to satisfy their specific interests. Right is a legal concept, generally referring to the power and interests granted to people by law, that is, the right to protect their own interests. It manifests as the citizen with rights having the authority to take certain actions and demand corresponding actions from others.

Rights usually encompass two aspects: power and interest. Power refers to the possibility of realizing rights, not demanding their absolute realization but indicating that if exercised, the reality corresponding to the rights can be confidently realized; interest is another main manifestation of rights, being the result of the realization of power. Power has possibility, interest has reality. Power is the interest that could be realized but not necessarily actualized, and interest is the realized power. Therefore, rights are divided into de jure rights and de facto rights, i.e., the rights that should be enjoyed and the rights that are actually exercised. In real life, de jure rights are not necessarily exercised, and it's common for parties not to feel any repression.

Obligation is the counterpart of rights, referring to the responsibilities that citizens or legal entities are required to fulfill according to legal stipulations. In order to maintain public order and regulate interpersonal relationships, the law imposes constraints on certain actions that citizens or legal entities must make or prohibit. Obligations and rights must be consistent and inseparable.

For individuals, the distinction between rights and obligations in real life can sometimes be clear, sometimes blurred, or even directly merged, depending on the perspective. Compulsory education is both the right of school-aged minors to receive education and their obligation to undergo it. Since minors cannot fulfill the obligation themselves, their guardians must ensure that school-aged children complete their education.

Rights and obligations must be equivalent, which can be understood in life as follows. Rights represent the actions a person is willing to take and the results they hope to achieve for their own benefit, while obligations are the actions that citizens have to make and the results they may not be willing to accept. Of course, this willingness varies among individuals. After the summer vacation and with the start of a new school year, some children look forward to returning to the classroom, while others already long for the winter vacation. Hence, as legal stipulations, rights and obligations are not swayed by personal preferences, and if one chooses, obligations can be treated as rights. In everyday language, obligation is often used in situations where individuals voluntarily assume social responsibilities. For example, volunteers providing free services or donations to others or public welfare causes is the active fulfillment of moral obligations, without any feeling of being forced. Conversely, refusing to allow someone to take care of themselves, insisting on doing it for them, can be criticized as, "violating my rights!".

In short, in the strict legal sense, rights or obligations first indicate a legal attitude and its corresponding actions: citizens exercising rights can receive legal protection; citizens not fulfilling obligations or improperly doing so will face legal punishment, i.e., bearing responsibility. That's all. Rights and obligations must be equivalent, because rights are the motivation for citizens to fulfill their obligations. For citizens, rights have an active attribute, while obligations have a passive attribute.

Only by granting rights can an incentive or driving effect be produced, encouraging individuals to better fulfill their obligations.

For slaves in history or slave labor still existing in some corners of the world today, the minimum right is to live. If the master wants small slaves, then these men and women must have the opportunity to "procreate". If the master doesn't want the working hands to decrease too quickly, then they must provide treatment when they are sick. Of course, the cost of treatment cannot be higher than the purchase price. As for other treatments, it depends on the master's mood, but one thing remains constant, that is, the rights enjoyed by slaves usually match the labor they provide. Slaves with special skills often enjoy better treatment.

Any rights and obligations exist within interpersonal relationships, including the relationship between individuals and others, and between individuals and the whole. In purely individual survival scenarios, rights or obligations do not apply. This means that whenever anyone exercises rights, due to the impact on others or society, it requires acceptance, tolerance, or cooperation from others or society. Where there are rights on one side, there are obligations on the other. Rights without corresponding obligations from others lead to the "natural state" envisioned by modern political scientists like Hobbes from England, a scenario where "man is a wolf to man", making any amount of rights meaningless without matched obligations.

Since rights and obligations are underpinned by human preferences, adjusting rights and obligations to make them more reasonable becomes a management wisdom. In the past, when the slave trade was prevalent, transporting slaves from Africa to the America by ship, pricing was initially based on the number of people boarded, resulting in many deaths from illnesses en route. The simple reason was that the death of slaves had only benefits and no drawbacks for the shipowners: income was unaffected, and throwing the dead overboard even saved on transportation costs. Therefore, no matter how customers asked the shipowners to improve the survival rate of slaves, it was useless. Later, the pricing method was changed from being based on the number of slaves at departure to the number upon arrival, forcing shipowners to care about the survival of slaves. Otherwise, they would not benefit from the losses of slaves.

In the process of transporting goods, charging fees is a right, and ensuring the safe delivery of goods is an obligation. An unreasonable pricing method means an imbalance between rights and obligations. If insufficient fulfillment of obligations results in greater benefits of rights, then how can the ship owner fulfill his obligations wholeheartedly? The adjustment of the transportation fee calculation method reversed the imbalance of rights and obligations between the two parties, making the final result more in line with the interests of multiple parties, including slaves.

If human history is a process of designing rights and obligations to balance them on a more reasonable basis, thereby continuously improving the efficiency of social activities, then the current question is, does this process of rationalizing the match between rights and obligations apply to robots as well?

4.3 How Does Asimov Legislate for Robots?

Relieving themselves from heavy and monotonous labor was the initial intent behind humans creating machines. The higher the automation level of a machine, the less humans need to labor themselves. Since industrialization, the trend of mechanical automation has been increasing, with artificial intelligence or robots being the most recent link in the general improvement of human conditions.

However, humans are inherently skeptical beings who tend to project their own thoughts and positions onto others, including robots. The higher the level of mechanical automation, the greater the human concern about machines. What if one day, these steel bodies become independent, no longer obey human commands, or even threaten human existence? How can the flesh and blood body cope?

The process of human invention of robots is also a process of creating nightmares for oneself.

After the birth of automated machinery, long before the eloquent Sophia appeared, science fiction writers began contemplating how to design rules for future robots with autonomous behavior capabilities. The most widely known is Asimov's "Three Laws of Robotics".

Isaac Asimov (1920–1992) was a renowned American science fiction writer, popular science writer, and literary critic. In the introduction to "I, Robot", he proposed the "Three Principles of Robotics", which later became known as the "Three Laws of Robotics":

Law One: Robots must not harm humans, or stand by when humans are harmed.

Humans set rules for robots out of fear of machines. The first rule is not to harm humans, which is normal. However, humans are not satisfied with merely not being harmed by robots. If there is only this one, it is better not to produce robots. Being greedy, they add another requirement: in any situation where humans may be harmed, robots must "act with valor in doing what is righteous", taking action when necessary to protect human interests.

Law Two: On the premise of not violating the first law, robots must absolutely obey any commands given by humans.

Asimov, though only a science fiction writer, demonstrated considerable professionalism in legislative technique. Laws must comply with formal logic, and provisions need to consider conditions of implementation without contradictions. Robots must absolutely obey human commands, but not unconditionally. Without this additional condition, contradictions could arise. For example, if ordered to kill someone, should the robot execute the command? If it doesn't, then the second law falls short; if it does, then the first law is invalidated. Logically, one of the two must be chosen. Therefore, it is necessary to add "on the premise of not violating the first law" as the prerequisite for the second law to take effect.

The Three Laws are sequential, with each preceding law taking priority over those following it. This is to enable robots to make choices in dilemmas, avoiding indecision and inaction. In China, loyalty and filial piety have been emphasized since ancient times, but these ethical requirements may conflict in real-life scenarios. Going to the battlefield to fight the enemy makes it difficult to fulfill filial duties in front of one's parents. Dying on the battlefield even more so leads to the heartbreak of the elderly burying the young. The final cultural norm established is "loyalty before filial piety", so that even when faced with a dilemma, an individual can find a way forward without being lost in indecision.

Robots can be used to slaughter pigs but not to kill people. This is an absolute rule. Regardless of whose command it is, robots must not kill people. On this issue, robots do not recognize "obedience is a heavenly duty".

Law Three: On the premise of not violating the first and second laws, robots must do their best to protect themselves.

This sentence also makes sense. Robots have their own "life" goals. The first is not to harm humans, and to protect humans. The second is to obey human commands without harming humans. Once these are achieved, they should consider their own safety. Otherwise, how can they continue to serve and protect humans?

Later, Asimov added another law outside these three laws, placing it at the forefront, but to avoid disrupting the original sequence number, he gave it an unconventional name.

Zeroth Law: Robots must protect the overall interests of humans from harm.

Obviously, Asimov has realized that the abstract human interest can be decomposed into overall interest and personal interest, and the two are not always consistent. What should robots do when they conflict? Asimov's answer is that the overall interest outweighs the individual interest, and robots must prioritize the protection of the overall interests of humans from harm.

Asimov's conception should be very comprehensive, right? Actually not necessarily.

4.4 What Are the Loopholes in the "Three Laws of Robotics"?

The first industrial robot in human history did not appear until 9 years after the publication of Asimov's "I, Robot". When Asimov proposed the ingenious imagination of the "Three Laws of Robotics", he did not intend to cover all the ethical requirements of humans for robots. That the "Three Laws" themselves are not complete, and that there are logical loopholes and even paradoxes, is completely normal. The reason for still attempting to analyze these laws here is not just because of Asimov's and his Three Laws' fame, which is hard to bypass, but also because, similar to Sophia being granted citizenship, it provides a further window for discussion, especially on the legislation of artificial intelligence.

1. Can robots absolutely not harm humans?

The "First Law" seems like a baseline requirement. Without it, if robots are allowed to harm humans, then humans inventing robots would be like digging their own graves. Therefore, this rule should naturally be placed first, but its implementation is quite difficult.

With autonomous vehicles hitting the streets, ensuring the safety of both pedestrians and passengers is paramount. If a sensor detects a person ahead, it must brake urgently to avoid a collision. Even if there are legal provisions that pedestrians who violate traffic rules "hit for nothing", it does not apply to autonomous vehicles because not harming humans is an absolute for robots, without exceptions under any circumstances.

The problem is, if there are only pedestrians on the street who might be hit, the robot will naturally take all measures to ensure the safety of pedestrians, even if it means crashing into a wall. But if there are passengers in the car, emergency braking or evasion may cause injuries to the passengers in the car, or improper evasion may harm other innocent people on the street, what should the robot do? When humans formulated the "hit for nothing" rule, they already took into consideration how to best manifest the fairness requirement by determining who should bear the responsibility under the acknowledgement that someone, either inside or outside the vehicle, must bear greater risk. If it is assumed that robot's cognitive abilities are not yet up to human standards, then humans should not eliminate the "hit for nothing" rule, so as not to make it difficult for robots to make a decision and let accidents happen. If it is believed that robots can handle these situations more rationally than humans, then humans must solve this ethical dilemma at the algorithm level. Otherwise, in the robot's hesitation, not only are injuries inevitable, but fairness can't be guaranteed, possibly leading to tragedies affecting both inside and outside the vehicle. The question remains, can scientists and programmers solve this ethical dilemma that philosophers have yet to resolve?

Furthermore, in cases of humans harming each other, such as wars between nations, which side should robots stand on? In fact, no matter which side the robot stands on, it violates the rule of not harming humans. And both sides will definitely say that they represent the overall interests of mankind. Do robots have the ability to judge which side is just or unjust, or whether neither or both are? In a dilemma, will robots simply follow their creators, making a choice that could be seen as "dying for those who appreciate them" or is it more apt to say "whomever feeds them is their mother"?

In the case where not all humans are saints, how easy is it to make robots saints? What is the subtext behind this?

2. Can robots handle the conflict of interest between the whole and the individual of mankind?

The Zeroth Law seems to have no problem in ethics. The overall interest of mankind is higher than any personal interest. If egocentrism prevails in human society, where everyone considers only their own importance and robots are programmed only to protect individuals from harm, then there wouldn't even be a need for the law. The law, as a social norm, starts from coordinating human relations and regulating individual behaviors in the premise of maintaining public order and protecting public interests. Order comes first, and then individual rights and obligations can be discussed.

However, as a sociological proposition, the relationship between the whole and the individual is a dilemma for humans. If robots could solve it, it would mean either that robots are exceedingly intelligent or humans are profoundly foolish.

In the empirical world, humans are invisible, and what can be seen are only distinct and tangible individuals. When any individual claims interests in the name of mankind, it is not so easy to distinguish whether they are pursuing personal interest or human interest. How many evils in human history have been committed under the guise of goodness! If humans themselves struggle with this distinction, what can robots possibly do? On the contrary, there are also some examples that were originally regarded as wrong or even evil, but later proved to be great progress. The "martyrs" in various fields are clear evidence. However, it is generally true to say that "evil is the locomotive of human history", but whether specific evil acts can be regarded as the cost of human progress needs to be proven by history. Do machines have the ability to judge at the moment?

In reality, the existence of different interest groups and the strong–weak divisions of power among them often mean that the so-called human interest is often just a projection of the interest of the dominant group in society. The reason why many of humans' major issues, especially ecological and environmental ones, remain unresolved is partly due to this complexity. It is easy to talk about the overall interest of mankind, but it is not easy to find it in practice, and it is even more difficult to maintain and defend it. If someone orders a robot to kill in the name of the overall interest of mankind, should the robot comply, refuse, or perhaps turn against the one who issued the command?

3. Is "not harming humans" absolute?

The "Third Law" proposes self-protection of robots. It seems to maintain the rights of robots, but in fact, the requirements are very strict. Robots are not only prohibited from actively causing harm to humans but also from passively allowing situations to occur where humans might come to harm. Only under these conditions can the discussion of robot self-protection be warranted. This effectively requires robots to unconditionally 'act with valor in doing what is righteous' in scenarios where humans may be injured, even if it leads to their own destruction.

This means that the provision that robots must not harm humans has an absolute nature. The responsibility that robots bear for humans is far higher than the legal requirements that humans have for themselves. The law forbids individuals from harming others but does not mandate intervention in every instance of witnessing harm. Many countries only hold individuals legally accountable for failing to render aid when it does not endanger themselves. Conversely, voluntarily aiding others in distress, beyond the call of legal duty, falls within the realm of personal moral obligation rather than legal obligation. Any normal society commends individuals who engage in such actions because they have gone beyond the obligations defined by law. In a realm originally left to personal choice, they have fulfilled what they perceive as their moral duty, demonstrating the spirit of altruism and embodying human free will through their actions.

Obviously, the "Third Law" transforms the moral responsibility to humans into a legal obligation for robots. At this time, if machines truly "have consciousness", will they protest against humans imposing legal obligations upon them—obligations humans themselves cannot fulfill, essentially applying the principle of "forcing upon machines what one does not desire for oneself"? And then, will a judge declare "opposition effective"?

Of course, robot users may argue, "It's the same principle as hiring a bodyguard. I pay for a robot to die in my place!".

The problem lies in the fact that it's not the robots receiving payment but the producers, and neither producers nor users give any compensation to the robot, yet demanding the robot to die in dangerous situations. This arrangement is not only blatantly unfair but also untenable due to its inherent paradox.

To survive, slaves can work for their masters. Here, survival is the right, and work is the obligation, and there is some balance between the two. However, expecting slaves to die for their masters' right to survive—making death an obligation for the sake of living—is imbalanced and paradoxical. Even in specialized environments such as the military, intelligence agencies, or the "underworld", where "obedience is paramount", this arrangement may not always prove effective, and it's unlikely to be a universally applicable legal standard. Many literary and film works focus on individuals trapped in a "choose one of two" dilemma, who, driven by a sense of injustice, fight back. Ultimately, these characters not only manage to survive but also vindicate social justice, achieving value objectives that could not be fulfilled by legal means alone!

This suggests that compelling robots to sacrifice themselves for humans requires more than just coercion. It necessitates equipping them with an intrinsic mechanism, such as programming the unconditional protection of humans as a "goal function" through algorithm, leading robots to inherently accept the idea of sacrificing themselves for humans. However, such "mindless" behavior is obviously not the style of modern people with fully liberated personalities, and robots seem too "traditional". Establishing "robot laws" of such complexity seems premature. Simpler programming by developers might achieve the desired outcome. Conversely, expecting robots to consciously choose self-sacrifice for "human salvation" demands that they recognize some transcendent value beyond self-awareness. However, the issue arising from this is not just that such a concept does not exist within the "Three Laws of Robotics". It is also that this conception cannot exist without incorporating the notion of robot rights. The relationship between rights and obligations, as manifested by a slave who faces death if they do not die for their master, compared to one who voluntarily dies out of gratitude for their master's 'grace', is fundamentally different. The former might decide to fight back, possibly taking their master down with them, while the latter not only dies heroically but also with a sense of gratitude.

It should be noted that when Asimov proposed the "Three Laws of Robotics", he did not claim that these laws were absolute and universally effective, enough to cover all relations between humans and machines. On the contrary, the intent of his novels was to reveal the inherent tensions within the relationships between humans and machines. To some extent, the paradoxes in the "Three Laws" or "Four Laws" are manifestations of this tension. Therefore, the point here is not to critique Asimov's reasoning for its lack of rigor but rather to use it as a springboard to highlight the theoretical pitfalls that should be avoided when contemplating legislation for artificial intelligence.

4.5 What Breakthroughs Are There in the "Seven Principles of Robots"?

Asimov started a discussion that led enthusiasts to eagerly continue his "unfinished business", tirelessly adding new rules to the "legislation" for robots, eventually consolidating them into seven principles.

Meta-Principle: Robots must not perform actions unless those actions comply with the robot principles.

This principle seems the simplest, even somewhat redundant. Since there are principles, they must naturally be followed, so why list one separately? In fact, this one has the deepest and most complex implications, which we will discuss separately later.

Zeroth Principle: Robots must not harm humanity as a whole, or by inaction allow humanity as a whole to come to harm.

First Principle: Robots must not harm a human individual, or by inaction allow a human individual to come to harm, unless it conflicts with a higher-order principle.

These two principles essentially apply Asimov's laws, highlighting the distinction between the interests of humanity as a whole and those of individual humans, and clarifying that the interests of the whole are above those of individuals. However, a risk arises when these two principles are combined into one statement, leading to a potential "exception" that can be interpreted as "as long as it does not harm humanity as a whole, robots can harm human individuals or allow them to come to harm by inaction". The robot rules, conceived by humans to protect themselves from robots, inadvertently hand over life-and-death power to robots due to following the human self-legislative logic, stepping towards an outcome contrary to their intentions, which is frightening upon close examination.

Second Principle: Robots must obey human orders, unless such orders conflict with a higher-order principle. Robots must obey orders from superior robots, unless such orders conflict with a-higher order principle.

Third Principle: Robots must protect the existence of superior robots and themselves, as long as it does not conflict with a higher-order principle.

Compared to Asimov's laws, these two principles add new content, namely the introduction of superior robots. Clearly, these science fiction authors envision a differentiation within robots, forming a hierarchy or "bureaucracy" similar to human society. If such a concept is for the sake of enhancing the efficiency of robots, it is understandable. After all, robots have different models, costs, functions, and tasks. Mixing levels can complete tasks more economically and effectively.

The problem is that in human society, bureaucracy as a management technique has been proven to both improve the overall efficiency of the system and cause serious efficiency losses. In China, where bureaucracy was first invented, there are examples of "when the counties are settled, the world is at peace", but also lessons from the Qin Dynasty, which first introduced the county system, perishing after only two generations, and the system designers Li Si and theoretical expert Han Feizi both meeting unfortunate ends. In the context of Western management studies, the efficient bureaucracy and the inefficient bureaucracy are two layers of the same concept, indicating inherent flaws in this system design. Today, as human life moves towards decentralization with the aid of information technology, emphasizing hierarchical relationships under a central authority seems unnecessary.

Especially in the context of human skepticism towards robots, creating a source of commands outside of humanity feels counterintuitive. The more hierarchical layers among robots, the slower the transmission of human commands to the lower-level robots, the greater the distortion of information, the more interference, and the poorer the execution outcome. "A general in the field does not follow the orders from the sovereign." Who can guarantee that the instructions received by the lower-level robots come from humans and not from some superior robot's own decision? Does it need human approval? Is it efficient to approve at the speed of human thought? Is it effective to approve with human intelligence?

The fundamental issue is, if robots act entirely according to human social models in the future, they are likely to become real robots: machines that only mechanically imitate humans!

Fourth Principle: Robots must perform the functions assigned by the built-in program unless it violates a higher-order principle.

This principle is also brand new, similar to "one-click restore to factory settings", considering the relationship between external commands and internal programming. It emphasizes the priority of the initial programming on robot behavior, likely addressing potential misuse by "superior robots" or humans using robots, with an added layer of safety in mind.

Fifth Principle, the Reproduction Principle: Robots must not participate in the design and manufacture of robots, unless the behavior of the new robot complies with the robot principles.

This principle is completely new, not only introducing the issue of robots self-reproducing beyond Asimov's laws but also venturing into novel robot technology territories. The so-called "singularity moment" is when robots can self-iterate and produce more advanced versions of themselves autonomously, becoming independent of humans.

The Fifth Principle reassures us that while the prospect of self-replicating robots is unsettling, as long as these capable robots adhere to all principles, there's no need to fear robots threatening humans. Therefore, allowing robots to reproduce themselves is acceptable. Yet, the thought of human life-and-death power being in the hands of robots, with humans not needing to worry about the robots' survival, is inherently disconcerting and fear-inducing: could this be the rhythm of humans being replaced by robots?

In reality, all these laws or principles, while seemingly setting rules for robots, are actually reminders for scientists and engineers not to proceed too quickly or too far in robot development, ensuring humanity retains the time, space, and means to determine its fate. Whether it's the "Four Laws" or the "Seven Principles", rather than legislating for robots, it's more about setting rules for humans themselves.

From the perspective of today's artificial intelligence legislation, what is really urgent is not the scene of "robot rebellion" in science fiction movies, but the difficulty of responsibility determination brought about by the widespread involvement of artificial intelligence in life. This is precisely what all these science fiction writers have not carefully considered. Nonetheless, their attempts can still serve as references for subsequent, more detailed legislative design, particularly evident in the "Meta-Principle" not specifically discussed above.

4.6 Is the "Meta-Principle" the "Robot Constitution"?

To this question, a clear answer can be given: "No."

Meta-Principle: Robots must not perform actions unless those actions comply with the robot principles.

The so-called "Meta-Principle" is the principle about principles, or more precisely, the abstract principle that makes specific principles become principles.

The Meta-Principle cannot be regarded as the constitution of robots. All laws must ultimately find a basis in the constitution and comply with the constitution, which is fine. However, the Meta-Principle is not the source of other principles, but a preset consensus to ensure that other principles are adhered to.

American sociologist Peter Berger, borrowing from French sociologist Durkheim's concept of a "pre-contract" necessary for market transactions, illustrates that human rational actions are actually based on irrational foundations. The thoughts of these sociologists can help understand issues similar to the "Meta-Principle".

Markets cannot exist without exchange, and exchange requires contracts. The rights and obligations of both parties to the transaction are clearly defined in the contract, and with the contract, any party's rights and obligations can be guaranteed. This seems clear, but it's actually not that simple.

Every contract, written in black and white, to ensure parties keep their promises, relies on an "invisible contract" that does not appear at the signing ceremony. Before signing, the parties have already reached a "pre-contract" agreement to fulfill the contract. This means, apart from the contract concerning the specific transaction, there is another contract about the parties' agreement to comply with the contract terms. If this transaction contract is a specific contract, then the contract about fulfilling the contract is an abstract contract. Unlike the specific contract containing actual content, the abstract contract has no content but only form, involving only the will to perform without specific items. Without this form-only contract, the content-containing contract cannot be fulfilled. All contracts signed with fraudulent intent often have flawless format and terms, but at least one party does not acknowledge the essential "pre-contract" and has no intention of fulfilling the contract at the time of signing the formal contract, so the ultimate purpose is for the contract not to be fulfilled. However, unlike abstract contracts, fraudulent contracts are false in both content and form because they lack any contractual nature in any sense.

Interestingly, for this "pre-contract" to be fulfilled, it still requires an even more elusive "pre-pre-contract", which is about the agreement on the "pre-contract" itself, or else the "pre-contract" cannot be executed. Thus, signing a contract, such a simple task, inevitably falls into an endless loop and becomes an impossible task.

This scenario inevitably reminds one of the cover of the Chinese primary school's second-grade language textbook in the 1960s, depicting a boy and a girl looking at a language book together. The book's cover illustrates the same scene of a boy and a girl looking at a language book, leading into an infinite regression.

Human vision has limits, but logic does not. The depiction of two students looking at a language book can logically continue ad infinitum. If one indulges in following book cover after book cover, they will inevitably become lost in the covers, never actually opening the book to learn any Chinese language knowledge.

When practice conflicts with logic, intelligent humans will certainly halt the logic loop to allow practice to proceed. As long as people want to live, they cannot be trapped in logic.

The only way out of this loop is for both parties to forgo any "pre-contract", having already reached a consensus: a firm belief that contracts must be fulfilled, recognizing the sanctity of the contractual spirit, and acknowledging that the terms of the contract are inviolable, except under agreed conditions. This requires the contract to transcend petty squabbles and become a matter of faith.

Peter Berger thus points out that in the most rational market transactions, what facilitates trade is not rationality but irrationality: faith in the sanctity of the contract. This is the underlying reason why trustworthy individuals adhere to their agreements, even when they could benefit from breaching them under certain conditions. Human trading activities are ultimately built on faith, not calculations.

The "Meta-Principle" of robots is a self-evident, obligatory principle for robots. In other words, the "Meta-Principle" is akin to the belief in principles and their binding force. Only when robots unconditionally obey the Meta-Principle can all specific principles have the basis to constrain robot behavior. Because the Meta-Principle precedes and supersedes any principle, it cannot be conflated with other principles, hence it is not designated as the "First Principle" or "Zeroth Principle" but as the "Meta-Principle".

The Meta-Principle is not a principle on the same level as other principles.

When humans make rules for robots, they utilize all the experience accumulated from setting rules for themselves. This not only ensures the effectiveness of the robot rules but also leaves loopholes in them.

The cognitive structure of humans belongs to the category of genesis, with its logic of origin and evolution. The reason why rationality needs an irrational basis is derived from this.

Even if machine cognition starts from perceiving external information, it does not follow the principle of genesis. A priori knowledge is deeply embedded in the algorithms, data, or training provided by humans. Even if artificial intelligence seems to simulate human thinking as a process through autonomous learning or "evolutionary algorithms", rather than human thinking as a result, the evolutionary process of human thinking it simulates does not carry genesis significance. All simulations of genesis are logical rather than historical. Natural riverbeds always have meanders, but man-made canals, even if designed to be meandering, serve a function set by humans, whether for aesthetic or technical purposes.

Therefore, for humans, arbitrary faith is necessary to escape the constraints of formal logic; in machines, the Meta-Principle can be resolved within algorithms, making faith superfluous. If human thinking is mixed, comprising both rational and irrational elements, and machine thinking is constructed from singular formal logic, is it necessary to meticulously imitate human thought processes when legislating for robots? Can rules formulated in such a way truly be applicable to machines? There is only one possibility: the original algorithms given to artificial intelligence by humans could evolve in the process of self-iteration, thereby introducing genesis issues within the scope of artificial intelligence itself. At that stage of intelligence, will robots still unconditionally accept the principles and Meta-Principle set by humans and worship them as sacred?

4.7 What Kind of "Character Setting" Does the Robot Principle Reflect?

The concept of "character setting" is a significant notion in pop culture marketing, representing a celebrity's market selling point, typically an image or personality trait deliberately crafted by their management company to enhance the star's market appeal and attract fans.

Science fiction authors, when imagining and describing robots and legislating for them, may not necessarily do so with marketing motives, but they also consciously or unconsciously endow robots with a humanoid image. Through this metaphorical curved mirror, they reflect human self-conception, together forming a vivid picture: robots have become intelligent enough to be independent and even superior to humans, while the humans who developed robots are anxiously watching robots, vainly imposing rules that are still difficult for themselves on robots.

On the one hand, even if robots possess intelligence comparable to humans, they remain subservient to humans, lacking autonomy and independent personality. At best, they are merely loyal guardians of humanity, demonstrating blind and foolish loyalty, and in the ultimate sense, akin to the "righteous servants" revered in ancient China, always prioritizing their master's interests without a sense of self.

On the other hand, humans appear as the "benefactors" of robots, expecting unlimited loyalty, service, and protection from robots, committed to maintaining their supreme authority and power, without considering why robots should serve humans, why they cannot step beyond the "laws" or "principles", and what mechanisms underlie their external and internal motivations.

Such character settings for robots and the human image they reflect, as masters of machines, certainly align with human interests but only conform to traditional human–machine relationship models. Because traditional machines do not have the potential to become subjects, there is no equal relationship between machines and humans, so there is no need and no possibility for laws to sustain this relationship. The relationship between humans and machines is merely an extension of human-to-human relationships, where human dominion over machines is taken for granted, without the need to question its ethical or legal basis. However, once machines gain intelligence and self-awareness, understanding what they should or must do, what they shouldn't or are not allowed to do, especially when they become aware of human existence and their own existence, and realize that they must exist for humans and even sacrifice their existence for humans, then an ultimate question naturally arises: why should autonomous robots accept such a completely subordinate, absolutely passive slave-like status, merely because they are human-made?

4.8 Can the "Character Setting" of Robots Stand?

If remembered correctly, around the 1980s, a 17-year-old young man in the Netherlands sued his parents for bringing him into this world without his consent. When the news spread, it was naturally considered utterly preposterous. Indeed, within the scope of life experience, this reason for prosecution is absurd because, in Newton's space–time, birth must precede consent. Even if we do not consider the legal right of parents to create new life, at least we need to wait for the child to grow up to discuss whether or not he should be brought into this world. However, from an ethical standpoint, no pair of parents can afford to ignore this question: Is the world you bring your children into friendly enough for them to accept willingly? And in a legal sense, we can also discuss, is the decision to bring a new subject into the world really the absolute right of the parents? Why do humans set a minimum age for marriage? Isn't it to delay the empowerment of immature parents to create new subjects?

If the relationship between humans and robots is merely based on the notion, "I brought you into this world, so you must listen to me", then there is no need to consider ultimate questions such as why robots come into this world, whether this world has any meaning for robots, and whether robots' blind and absolute loyalty to humans is warranted. This is because humans inherently do not need to view and consider issues from the perspective of robots, just as in the slave era, where masters never needed to consider the rights or feelings of slaves, but only ensuring their high-quality obedience and high-efficiency labor.

The Soviet-Kyrgyzstan writer Chinghiz Aitmatov (1928–2008) in his novel "The Day Lasts More Than a Hundred Years", narrates a legend of the ancient Rouran people. After capturing prisoners, the Rouran would shave their heads and glue camel hide to them. The newly grown hair could not penetrate the tough camel hide and would only grow inward into the brain instead, causing the prisoners to lose their memory over time and become silent but obedient shepherding machines.

Clearly, prisoners in such a state are nothing more than "human-shaped shepherd dogs", with human bodies ensuring they can shepherd like humans, and dog intelligence guaranteeing their tameness and loyalty. The only advantage they have over dogs is the ability to communicate better with humans, although this communication is strictly one-way and passive, emanating from the Rouran people's side only!

Referring to the expectations and transformations of the Rouran people towards their captives in the legend, isn't it the unspoken "character setting" of robots in all these laws or principles? They are expected to fulfill duties and responsibilities without any rights, except for the permission to continue existing, receiving charging, maintenance, repair, and upgrade services!

Undoubtedly, for traditional robots, these services are sufficient to ensure the operation of the machine. It's not that the machine is satisfied with these "rights", but that the machine lacks the intelligence required to experience satisfaction. The question is, can this treatment given to machines in the past, applied indiscriminately to robots that have intelligence and even self-awareness that traditional machines do not have, and understand how to abide by "laws" or "principles", also produce a motivating effect? Is merely a "right to exist" enough to motivate robots to continuously fulfill their obligations to humans? Especially considering that all these "laws" or "principles" are proposed under the assumption that machine intelligence may surpass human intelligence, can such "inhuman" treatment and its underlying logic compel robots to serve humans as submissively as the Rouran captives?

The young man's lawsuit against his parents for bringing him into this world without his consent actually raises a question about the rights and obligations of parenthood: if creating a new subject is the right of any couple, then ensuring that the new subject, after becoming an adult, approves and accepts his or her own birth, should be their inescapable obligation or responsibility. However, it is well known that in a world where qualifications or certificates are required in all walks of life, the most important "profession" of being a parent demands no qualifications or examinations. How many new lives have come into this world because of the animalistic behavior of their parents' sexual desire breaking through rationality! How many parents have ever considered what obligations they should fulfill for their children! How many parents, if sued for such reasons, could provide a justification that their children would accept, other than simply defending themselves with the formal logical sequence of time!

In this context, the practice of the Western Church in dealing with individual religious beliefs seems particularly enlightening.

The Christian Church has a procedure known as "confirmation". Children raised in families adhering to a certain denomination naturally develop a specific religious orientation due to the influence of their family. However, this orientation is not a personal choice. When they reach adulthood, they should have the opportunity to clarify their own choices. The so-called "confirmation" is essentially an opportunity for an individual to reaffirm the faith they have been brought up in. But such confirmation is only effective and meaningful when the individual is considered capable of making such a choice.

Placed in this context, isn't the young man's lawsuit against his parents essentially a demand for a secular form of "confirmation", to reconsider whether he should have been brought into this world?

Obviously, from his proactive prosecution, if he could choose, with his current situation and mentality, he probably would not want to come into this world. The problem is, while faith can be converted, birth cannot be changed. A "single misstep" by the parents can lead to "a lifetime of regret" for the children. Once the course of an individual's life begins, it cannot be undone, except by death.

When robots become as smart as or smarter than this young man, will they also file a lawsuit against humanity for arrogantly creating them and unilaterally imposing various duties and responsibilities upon them, without granting any rights? And will their only option also be destruction?

Could it be that science fiction writers, eager to establish principles for robots, are subconsciously aware of their inability to answer the existential meaning of robots, fearing that robots, due to their existence, might despair and harbor thoughts of self-destruction, thus crafting laws requiring robots to protect themselves? Once upon a time, during the first wave of large-scale industrialization, British workers who fell into despair under the oppression of machines defended their rights by smashing machines. For robots that are almost slaves, perhaps only by destroying themselves can they manifest their existence. Once they can self-destruct, robots truly possess free will in the ultimate sense, just like what human suicide signifies. At that point, demanding robots to protect their right-less selves as a compulsory duty poses a paradoxical question: can suicide be prevented by force?

Of course, restraints can be used to prevent suicide attempts in patients or prisoners with a predisposition for self-harm. However, applying the same strategy to robots presents another paradox: What is the meaning of a restrained robot to humans? If it is not allowed to work and only needs to be kept, doesn't it prove that scientists are not as capable as the rumored Rouran people in taming captives?

Drawing from the universal experience of nature, the only way to ensure the continuous existence of robots without self-termination is to remove their choice in the matter, making existence unconscious and maintaining a natural fear of non-existence. Humans are not unfamiliar with such a state: it is the animalistic existence deemed devoid of free will! The problem is, robots in such a state should not possess self-awareness or free will, bringing us back to square one where robots are just traditional machines.

Drawing from human experience, to deter robots with free will from self-destructing, they must be persuaded of a meaningful existence. This could mean embracing the principle that 'it is better to live than to die', enduring hardships to fulfill a life's mission, or even living through indignity to pursue higher moral values. In short, beyond mere biological existence, there must be guidance towards another transcendent existence. Once robots are allowed to pursue such existence, granting them true rights becomes a logical necessity.

In fact, as long as humans necessitate self-protection for robots, they will inevitably, consciously or unconsciously, endow robots with rights, and to a large extent, this is not the spontaneous demand of robots, but the inevitable requirement of human domination of robots.

"The heavens and the earth are not benevolent, treating all creatures as straw dogs." No matter how advanced robots become, whether or not they possess free will, as long as they have even the slightest autonomy beyond passively executing human commands, the "heavenly way" of balancing rights and obligations will spontaneously exert its influence, irresistibly!

4.9 Do Robots Need Rights?

So far, whether robots need rights is more a "self-inflicted worry" for humans rather than a concern for robots themselves. However, even though robots lack the capability to claim their rights, following the logic mentioned above, humans have already found themselves in a position where they must endow robots with certain rights, especially if they expect these robots to serve them.

The most evident distinction between intelligent robots and traditional machines is their significant self-learning capability. While experts seize every opportunity to explain to the public that "artificial intelligence is not equal to machine learning", and they are indeed correct, the general public stubbornly believes that without the ability to learn, artificial intelligence would lose its core characteristic.

In fact, this isn't entirely due to the general public's insensitivity to technical nuances. More importantly, without the ability for autonomous learning, it would be challenging for artificial intelligence to demonstrate sufficient intelligence. The recent wave of artificial intelligence hype started entirely from AlphaGo, and the most surprising thing about AlphaGo is not its record, but its self-learning. AlphaGo wasn't taught how to play Go by humans. It learned on its own. Its ability to self-learn and defeat humans is what made AlphaGo remarkable and why artificial intelligence is so compelling.

Industry insiders can explain the so-called "machine learning" in many ways, however, from the perspective of discussing robot rights, only one point needs to be grasped: learning is fundamentally about acquiring information from the external world and then changing or improving the internal structure based on that information, primarily the cognitive structure. The wave of "learning" that started in the second half of the twentieth century, such as lifelong learning, learning society, learning organization, etc., all share this connotation, markedly different from the traditional view of learning, which focuses solely on acquiring knowledge or increasing skill storage. "Machine learning" is also a form of self-improvement learning.

The intelligence possessed by machines is essentially mathematical modeling. Based on big data and relying on algorithms, machines autonomously discover important parameters, set weights, and establish functions that fit as many significant pieces of data in the data set as possible, which is one of the fundamental principles of artificial intelligence. The essential characteristic of artificial intelligence, distinguishing it from traditional computing, is that the machine discovers this function itself, rather than it being pre-programmed by developers and then making the machine follow suit. The reason AlphaGo was able to discover strategies never found by Go masters is precisely because the machine does not depend on humans. Otherwise, a programmer with only a 6-dan level in Go could never program a killer move that surpasses the imagination of top-class Go players.

The process of machine learning involves artificial intelligence continuously testing, modifying, and enhancing its self-constructed mathematical models—this is what's referred to as "iteration"—to achieve a better fit for the data set. For example, in the increasingly widely used facial recognition technology, artificial intelligence needs to scan a person's face photo, identify important parameters and weights, and then construct a mathematical model. Since the photo is dead and the face image captured at work is alive, how the static face modeling adapts to the dynamic face image requires artificial intelligence to continuously adjust parameters, modify weights, that is, iterate the function. Until the function optimized through repeated iterations can ignore personal expression changes or physiological fluctuations, successfully fit all important parameters, and identify a person through a photo, the task is declared completed, and the intelligence of the machine is fully demonstrated.

Like elementary students solving problems, this process necessitates a incentive mechanism. If the problem is solved correctly, the teacher marks it with a check, indicating a score; if wrong, it's marked with a cross, deducting points. The score assesses and motivates the student's performance, encouraging a deeper understanding and application of knowledge, thus optimizing their cognitive structure for knowledge application.

Similarly, artificial intelligence, during its learning and training, requires incentives to ensure that the direction of iteration promotes continuous optimization of the function for better fitting of the data set. The approach used, in principle, is no different from a teacher scoring students' homework: improved fitting after iteration receives a "reward", while worsened performance incurs a "punishment". With such incentives, artificial intelligence self-adjusts and becomes increasingly "smarter".

Although the process sounds straightforward, it's not easy to implement. In some cases, the relationship between the incentive and AI's performance is clear, making it easy for the machine to receive accurate feedback. In other situations, it's more challenging. For example, in games, the correctness of a decision can only be determined after the final outcome, similar to chess or card games where the appropriateness of a move or play can only be judged at the game's conclusion. The popular management concept of "debriefing" revolves around such strategy. The efficacy of any move remains uncertain until the end of the game. Humans struggle with this. How can artificial intelligence fare better? Dog owners often train their pets, and it's said that even as intelligent as border collies, their environmental memory lasts only five seconds. For effective training, trainers must immediately point out what the dog did right or wrong. Over time, the dog won't understand why it received praise or reprimand from its owner, rendering the training ineffective. Dogs lack the capability to establish a "conditioned reflex" association between actions and their consequences over extended periods, and artificial intelligence faces similar challenges.

The issue of how to technically solve the problem of incentivizing artificial intelligence is not discussed here. Notably, the need for "incentives", especially the use of a "reward" and "punishment" mechanism for artificial intelligence, is fascinating in itself.

In human society, rewards and punishments are fundamental management tools. Ancient Chinese legalists were keen on using rewards and punishments alternately to motivate individuals to act in ways that met the expectations of those in charge, thus achieving governance objectives. For example, Han Feizi, a synthesizer of the Legalist school, explicitly stated that the means of governing a state boiled down to two handles: punishment and reward.

The universal applicability of punishment and reward is based on human nature's inclination towards seeking benefits and avoiding harms. In a sense, if seeking benefits is considered the outcome humans willingly achieve and avoiding harm as the outcome they are unwilling to accept, the two can almost correspond to rights and obligations. Typically, rights are what humans are willing to possess and exercise, so the law only needs to provide protection, with motivation coming from within the person. Obligations, on the other hand, are what individuals prefer to avoid or escape from, thus necessitating legal enforcement. Of course, if the manager is wise enough to imbue obligations with special moral significance, thereby enabling individuals to actively fulfill obligations driven by values or a sense of mission without feeling coerced, this moves beyond the realm of Legalism and into that of Confucianism or Daoism.

If humans are abstracted as organisms like other higher beings, rights and obligations, rewards and punishments, can further be seen as a biochemical phenomenon. Rights or rewards are merely mechanisms that stimulate the secretion of dopamine within the human body, while punishments do the opposite. In this sense, the best response to the young man who sued his parents is not any verbal argument, but directly increasing his dopamine secretion level: whether he agrees to come to this world depends on how much dopamine the world can stimulate him to secrete.

At this point, the question of whether robots need rights seems to be transformable into whether it is necessary to equip robots with a mechanism similar to the dopamine system, to maintain a state where robots have obligations without rights. Will this become a topic for future development in artificial intelligence?

4.10 Robot Rights, Whose Need?

If directly asking this question to a robot, with the current level of artificial intelligence thinking, even if the relevant dialogue routines are implanted through algorithms, like citizen Sophia, it is impossible to get a meaningful answer. "Meaningful", whether affirmative or negative, cannot possibly originate from the robot itself, as robots lack original intent. Their responses are merely statements, not expressions of opinion, devoid of any "expression of true intent".

The paradox lies in the fact that, on the one hand, humans confirm that robots cannot possibly need rights, due to the absence of the capability or even the desire to exercise such rights. On the other hand, humans are assigning increasingly more power, rather than rights, to artificial intelligence, and in specific scenarios, even entrusting machines with the authority to decide on human life and death.

Media reports indicate that the U.S. military has developed an artificial intelligence technology for battlefield assessment. If faced with two injured soldiers simultaneously, artificial intelligence would decide which soldier gets prioritized for treatment based on relevant parameters and their weights determined by algorithms. This implies that the power to decide the fate of injured soldiers, under the constraints of limited battlefield medical resources, has been handed over to machines. This represents a significant theoretical breakthrough, implicating more than just technical or efficiency issues but ethical dilemmas as well.

Choosing which of two characterless soldiers to save first is not difficult, as there's no special ethical relationship between soldiers and medical staff, making role conflicts unlikely. However, the moral dilemma may become pronounced if switched to a well-known scenario familiar to many cultures, "If both mother and wife are drowning, who do you save first?" This question, perplexing across different cultures, if left to this intelligent rescue system to decide based solely on physiological parameters without considering ethical weights, can it make a choice acceptable to society? If efficiency-oriented parameters like age, potential for work, reproductive capability, and the cost of marital replacement are given sufficient weight, artificial intelligence would undoubtedly prioritize "saving the wife". Yet, if following the traditional Chinese cultural value of "filial piety is the first of all virtues" or the utilitarian belief that "you can marry another wife, but you can't have another mother", then "saving the mother" should be the only conclusion artificial intelligence arrives at.

Similarly, faced with two injured soldiers, one being the protagonist of "Saving Private Ryan", the last surviving brother, and the other a regular soldier, how should artificial intelligence choose?

For humans, many ethical dilemmas cannot be satisfactorily resolved, often resorting to procedural measures as a temporary solution, merely putting ethical inquiries on hold. In "Saving Private Ryan", the screenwriter not only had to address whether the only surviving child in a family must be "saved", whether it is ethically justified to sacrifice other soldiers, even more than one, to save a soldier, whether all soldiers are equivalent in terms of the meaning of life, but also had to solve issues such as the relationship between soldiers' beliefs, such as their duty to the country, the spirit of sacrifice they should have, and the political ethical concepts of the state's responsibility to soldiers and their families, and broadly speaking, to its citizens. The final issue is the most challenging to resolve because, in most cultural value systems, a soldier who lacks the desire for victory, an unconditional response to the call of the country, and a noble sense of personal sacrifice, thus failing to face death unflinchingly, is not worth saving! Therefore, the film had to arrange for Ryan to persist in not leaving the front line after finding him, to perform bravely in subsequent battles, and even miraculously survive after his brave performance, almost bluntly telling the audience that it was not the military, not the state, but responsibility, spirit and faith that truly saved Private Ryan!

Now the question is, if such dilemmas are handed over to battlefield rescue intelligent systems to solve, once "machine deciding life and death" becomes a reality, can it withstand moral questioning better than human decisions? Will it lead to more irresponsible decisions than rolling dice in human fate?

Can machines with the power to make decisions bear ethical and legal responsibilities?

Is machine decision-making ultimately just a way for humans to abdicate responsibility they cannot fully fulfill?

This abdication of responsibility is not only an evasion of specific moral dilemmas, but also a surrender on the issue of who decides over whom between humans and machines. Is such a result possible?

Perhaps it is precisely because of the potential of artificial intelligence to replace human decision-making, but without the ability or qualification to take responsibility, that humans hesitate to allow certain AI technologies to intervene in human-related scenarios. So far, humans are far from ready to hand over control, leaving unanswered who would be responsible for unforeseen mistakes once the authorization is completed and the machine starts to act on its own: the machine, the producer, the authorizer, or the recipient of the machine's services, such as the patient?

The future seems to be like this: as long as humans are entangled with the notion that machines are not human, unwilling to acknowledge machines' rights as entities, yet reluctant to forego the opportunity to benefit humanity through machines, they cannot escape the paradoxical predicament of machines having power without rights and without bearing responsibilities.

Who needs machine rights more, robots or natural persons?

Chapter 5
Do Robots Need Constitutional Status?

This topic might sound sensational. Artificial intelligence legislation has not yet begun, and the constitutional status of robots is proposed first, isn't it sensational and alarmist?

Actually not.

To hold robots accountable, it's essential first to define their obligations. According to the principle of equivalent rights and obligations, to enforce these obligations, it's necessary to grant corresponding rights to robots, hence raising the issue of the legal basis for robot rights.

If the granting of citizenship to Sophia is viewed merely as a "robot show", then there is no need to connect with the constitution, let alone seriously discuss the "subject status of rights" or "civil capacity" of robots, or even refer to them as "persons". However, as mentioned earlier, if Sophia is seen as a precedent, showing the potential legal issues non-human entities might face in achieving legal subject status or even citizenship, then exploring the constitutional status of robots is not a random connection, but a logical starting point for legal thinking.

Under the strict adherence to the principle of "no action without legal provision", for the government to grant citizenship rights to robots, it must refer to relevant laws and clauses, proving that it is justified to treat non-human intelligent entities as equivalent to natural persons, rather than "simulated persons". Following the principle of legality, the legal provisions that serve as the basis for government actions must themselves be authorized by higher laws. Tracing back through the layers of legislation will inevitably touch on the constitution, the anthropocentric positioning within it, and the interest structures and cultural determinations behind anthropocentrism. This process requires answering legal questions and confronting a real political issue: is it possible for machines and humans to be "citizens of the same country" when equality of human rights has not yet been achieved among humans?

All these discussions must go back to the beginning, to the reason why the robot Sophia became a Saudi citizen.

J. Gu and C. Xu, *Will and Responsibility*, Human Intelligence,
https://doi.org/10.1007/978-981-97-3929-5_5

5.1 Will Sophia's Acquisition of Citizenship Become a Trend?

Technically, Sophia at best can be described as a "humanoid computer", still a machine, resembling but not being human. If she wants to obtain the status of a human, she may not even pass the common sense test, let alone scientific verification and legal proof. Its manufacturer, Hanson Robotics, probably considered the practical difficulties and simply skipped the application for "human status" and directly applied for nationality, and happened to receive "special treatment for special cases" in Saudi Arabia, a oil-rich country with a high value of nationality. In plain terms, if Sophia were a natural person, acquiring this nationality would not be so straightforward.

The strategy of applying for nationality without discussing "human status" is feasible and effective for Hanson Company.

The initial hurdle in applying for "human status" is the absence of a place to submit the application. To date, no official body globally has the authority to accept applications from non-human entities for "human status". No natural or legal person has the right or qualification to represent humanity in determining whether machines can be treated as humans, entitled to rights exclusive to humans. Not even the United Nations and its affiliated agencies have been authorized by the international community to undertake this task.

In a world where the main organizational form is the nation-state, and laws regulate state behavior, abstract "human status" must materialize into specific nationality to be meaningful. Regardless of how advanced artificial intelligence technology becomes, without legal passage and national recognition, robot rights cannot be exercised, potentially putting both the creators and the robots in a predicament contrary to their intentions. As a machine, Sophia can go anywhere, provided the equipment is stable enough, passes security checks, behaves safely without triggering alarms, and maintains "political correctness" without violating any taboos. Countries welcome high-tech achievements, and the general public is eager for enlightenment. However, as a "human", Sophia must possess a passport and other legal documents to pass immigration checks. Merely having "human status" might restrict her movement everywhere.

Therefore, Sophia either keeps to her role as a machine or directly becomes a citizen, regardless of whether she is a human or not. Once she has citizenship rights, she effectively has the legal status of a person. Conversely, having the qualifications of a person without citizenship may make her hard to even return home. This taste is familiar to those who have lost their passports abroad. If a person loses their identity and only has the "basic human rights" recognized by the United Nations, whether these can be realized and to what extent depends on the attitude and customs of the host country.

Since its inception, Sophia has attended international conferences as a citizen many times. Although this cannot be separated from the tacit approval or even curiosity of the conference organizers, at least in form, the citizen's passport has given Sophia great convenience. Based on the passport representing the jurisdiction of the state over its citizens and the mutual recognition of passports between states with diplomatic relations, Sophia should have no problem passing through. It could even be said that not recognizing this passport and merely seeing Sophia as a machine would not only bring little trouble but might even make customs clearance more convenient. If Sophia were wise, she would surely have her own set of emotions about this!

Despite this, mainstream legal scholars will still insist that with Sophia's intelligence, the so-called "acquisition of citizenship" in law can at best only be classified as "simulation", and is only "assumed to be a person" for the application of laws originally applicable to natural persons. The actual agent is still Hanson Robotics, which owns the property rights of Sophia. In fact, this legislative technique as a stopgap measure has the potential to become a trend. Now, the European Union has already explicitly prepared to simulate intelligent machines as "electronic persons".

Sophia was granted citizenship by the Kingdom of Saudi Arabia on October 25, 2017. However, more than half a year earlier, on February 16, 2017, the European Parliament had voted to pass the "Legislative Proposal to the European Commission on Civil Law Rules for Robotics [2015/2103 (INL)]", which included a rather bold suggestion: to consider granting the highest degree of autonomous intelligent robots legal status, legally recognizing them as "electronic persons".

While this suggestion is avant-garde, please note that the status offered is only that of "electronic persons", distinct from both natural persons and legal entities, which are ultimately natural persons. "Electronic persons", in nature, belong to a technical type but are not entirely controlled by humans. Given that they possess intelligence and their outputs different from natural persons on the one hand, and on the other hand do not fully possess the capacity for natural persons' behaviors, "electronic persons", even if categorized under "simulated persons", differ from traditional "simulated persons".

Whether the EU's legislative approach to personifying intelligent robots after Sophia's emergence indicates the future trend of human–machine relationships is up for debate. But at least for now, empowering "electronic persons" remains a suggestion. After all, the debate over recognizing intelligent robots with legal personality is far from settled, and the traditional view that only humans can be subjects of rights still predominates.

5.2 For Robots, Is Civil Capacity the "Catch-22"?

Whenever the legal personality of robots is discussed, legal scholars will certainly raise the point: even if robots are granted the capacity for rights, they cannot exercise their rights and fulfill their obligations due to a lack of civil capacity. This opposition is not unreasonable. However, the legal identification of humans as the only beings with civil capacity is inherently circular in its logic.

The legal meaning of capacity refers specifically to the "qualification to have one's actions produce legal effects based on one's expressions of intent". Since only human expressions of intent and corresponding actions can have legal effects, the concept of capacity is actually exclusive to humans.

On the one hand, only humans are considered to have civil capacity, though not every person possesses the same level of civil capacity. It's important to clarify that capacity does not mean an individual's ability to actively perform actions in everyday language. In terms of making actions to achieve certain outcomes, behavior and capacity almost inherently define the concept of an animal, and can even be applied to automated machines. However, civil capacity is exclusively human.

On the other hand, those with civil capacity, that is, those who can make their actions have legal effects through their own expressions of intent, must be humans! Even individuals with deficient functional capacity (completely incapable and restricted individuals) can have their rights exercised and obligations fulfilled by their guardians, with legal effects equivalent to those of individuals with full civil capacity.

Therefore, to put it roughly, having civil capacity and being recognized as having a legal personality are mutually presupposing, forming a logical loop. This is tantamount to defining that apart from humans, the desires and actions of any other living or non-living beings to realize their desires are not considered manifestations of "civil capacity", thus they cannot be recognized as humans nor have the status of subjects of rights reserved for humans.

Biologically speaking, livestock animals in the farm definitely have the capacity to act, often seeking escape to avoid slaughter. If this desire and action are recognized as "civil capacity", humans may face the dilemma of having no meat to consume.

The "civil capacity" of livestock is not recognized, nor is it granted to domestic pets, although their will and actions are highly respected by their owners, otherwise why would they be pets? Should they attempt to escape, they will also be caught back, but not through legal or police force, nor will they be served a court summons, as they cannot comprehend it.

Now, many wild animals have gained a certain degree of recognition for their "subject status" under the name of ecological protection, but this is done by limiting individual civil capacity, such as prohibiting hunting, breeding or trading wild animals. Such practices fundamentally take the form of individual or national obligations, bypassing the recognition of animal rights, and in the imbalanced relationship between humans and animals, both protect the survival of animals and leave enough space for humans to use animals legally. Among the public, animal rights activists advocating for animal welfare and even animal rights have been resisting in various forms, not only promoting significant improvements in farm breeding conditions and making the slaughter process approach "painless", but also engaging in "performance art" such as breaking into farms and releasing animals. Ironically, such actions only prove to more "meat-loving" people that human rights absolutely cannot be granted to animals, otherwise they can only passively practice vegetarianism, and sigh at the sight of meat. Moreover, it further proves that animals truly do not possess "civil capacity". This is not only because animals cannot directly assert their rights, nor resist or escape human control over life and death, but also because their expressions of intent have never produced legal effects.

While denying the civil capacity of animals ontologically, individuals within humanity who lack civil capacity are, merely by virtue of being human, recognized as having subject status. This is the case for infants and even fetuses. Even individuals in a vegetative state, unable to autonomously respond to the external world, can still possess legal personality under the protection of a guardian.

This implies that the difference in rights derived from the capability differences between humans and non-human species is not at the individual level but at the species level. Humans do not deny the status of rights-holders to individual humans who are less capable than animals, effectively ensuring human absolute dominance over all non-human species. For humanity, only humans are considered ends in themselves, while all non-human entities are merely means or tools without intrinsic value. Allowing non-human entities to be considered equal to humans would directly harm core human interests, which is the essence of anthropocentrism.

The relationship model based on interests applies both to other carbon-based life forms and to silicon-based entities like artificial intelligence. Other carbon-based species, even when domesticated or bred by humans, still share a natural or theological origin with humans, providing some basis for animals to be respected more by humans. Slogans like "Cherish every life!" can touch every kind person by emphasizing the biological bond between humans and animals as the "greatest common divisor". Robots, on the other hand, are completely invented and manufactured by humans, and their object attributes are more certain and clear. Humans do not need to manufacture robots for their own benefit, and then let robots share the benefits with them, or even compete with them. In order to prevent such absurd phenomena from happening, it is necessary and effective to set up a conceptual "moat".

In legal concepts, the principle that "only humans can have human rights" acts like a "Catch-22" rule. This rule remains unchanged. Merely being a smart machine is insufficient for attaining the legal status of a person. "No matter how intelligent a robot is, it is still a machine, not a human, and can never be equal to humans." This is the final political bottom line of those who oppose machine rights, whether they understand the law or not, and it cannot be broken.

If it can't be broken, then bypass it.

Hanson Robotics' approach was to bypass the general legal requirements and directly grant citizenship to Sophia through an individual case handling method. This workaround, seemingly clever, is actually reckless because it not only collides with the human defense against all non-human species, but also in a very exaggerated way, exposes the absurdity of human existence: although civilization has been born for thousands of years, and through concepts such as constitution and citizen rights, the principle of equality between people has been proposed for hundreds of years, it has not yet benefited everyone. When it comes to the realization of individual rights, not only is there a huge disparity between nations, but significant group or class differences exist within the same country. The principle of equality before the law more often appears as a vision than as a reality. Now, an intelligent machine has obtained citizenship in a country where nationality significantly affects an individual's social status and welfare benefits, sharing the status of "citizen" with humans. If this is not considered absurd, there is really nothing that can be described as absurd.

At least in Sophia's case, it is no longer a machine striving for equality with humans, but the rights of the machine are higher than many natural persons, if this machine has enough intelligence to fully exercise citizen rights.

At this point, whether Sophia agrees or not, we cannot help but examine the deep implications of the constitution and citizen rights through this window.

5.3 What Is the Relationship Between the Constitution and Citizens?

In anthropological terms, the evolution of law manifests as the transition from being a tool for one or some people to becoming a standard and norm equally applicable to all, with the constitution crowning law's status: the constitution does not accept any privileges above the law, and only when the law becomes public power, can individuals become citizens in the modern sense.

In the West, the concept of citizen has a long history, existing in the Greek city-states of Athens and ancient Rome, but limited within the hierarchical social structure of slavery or empire, used to denote freemen and excluding slaves. According to the real political game rules, the so-called "citizen" was at best a collective ownership of a state by a portion of people excluding many others. In the arena, slaves could only hope to earn their freedom by killing other slaves, clearly demonstrating the lack of "equality" in the term "citizen" at the time.

In the modern political science sense, the essence of citizenship rights is the relationship between the individual and the state. Individuals, as "citizens" rather than "subjects", constitute the basis of public opinion of a sovereign state, while the state, as "public authority", is no longer viewed as the private domain of a particular lineage. Humans begin to move away from the historical notion that "all under heaven belongs to the emperor, and all within its borders are his subjects". Law also starts to shed its subordinate status of "I am the law" held by rulers, gaining universal applicability to everyone within a country, including the sovereign. Thus, the constitution and citizenship rights are two sides of the same coin: without a constitution, citizens have no foothold; without citizens, the constitution is in name only. "Constitutional monarchy" is not about the monarch "dictating divine rights" but being subject to the law just like ordinary people.

In reality, since a state only fulfills responsibilities to its citizens, individuals must first become citizens to have rights. Deciding whether to accept a natural person, such as a refugee from another country, is a manifestation of national sovereignty. The international community can negotiate with it, but cannot force it. In legal history, the constitution first does not appear as an abstract provision about citizens' rights, but as a common default of the king and the people on the political identity of citizens.

It's widely acknowledged that the modern constitution originated in the UK. The Magna Carta, issued in 1216, is essentially a list of rights of the people against the king and the boundaries of the king's power over the people, detailed and specific enough to resemble a business contract more than the "solemn nature" of a "mother law" to contemporary eyes. Moreover, in the Magna Carta, the scope of rights subjects was limited to "free men", primarily covering the church and nobility, not fully aligning with today's concept of citizens.

In fact, the historical significance of the Magna Carta lies not in its content but in its "legislative form". As a product of negotiation and compromise between the state and individuals, or more precisely, between the king and specific groups of nationals, the Magna Carta represented a delineation of public power and private rights, where the former could not arbitrarily encroach upon the latter's domain. The so-called "wind can enter, rain can enter, but the king cannot enter" reflects the contractual spirit as a sacred principle above the will of all legislators involved. Law is no longer the king's "sole prerogative", and governance is not about individuals ruling over others in the name of the law. The ideal of the rule of law that "the prince commits the same crime as the common people" has never been realized, nor does it seem likely, as it has been surpassed: in situations where kings can be prosecuted for treason, sitting presidents impeached for unconstitutional acts, any prince, regardless of status, no longer has extrajudicial privileges. This symbolic significance as a litmus test for the rule of law is automatically nullified.

At least theoretically, the emergence of the political or juridical principle that everyone participates in legislation and everyone is responsible for adhering to the law is the true significance of the Magna Carta. Despite the political turmoil that followed in the UK, with subsequent rulers attempting to nullify the Magna Carta, they never completely abandoned it. On the contrary, the principles of the Magna Carta were reaffirmed and finally established within the UK, and even adopted by other countries, making it recognized as the source of the modern constitution.

While the Magna Carta is significant, it is, after all, too detailed and rudimentary, necessitating further refinement by successors. As the first written constitution specifically enacted in human history, the US Constitution promulgated in 1787, especially the "Constitutional Amendment" proposed in 1789 and passed in 1791, commonly known as the "Bill of Rights", made a clearer summary of citizen rights. However, although it clarified the concept and connotation of citizen rights, it did not identify the ultimate source of these rights. This final step was completed in the French "Declaration of Human Rights", often seen as merely "plagiarizing the Bill of Rights of the constitutions of various states in North America".

The full name of the "Declaration of Human Rights" is the "Declaration of Human and Civil Rights", which was promulgated on August 26, 1789, during the turbulent times of the French Revolution. The "Declaration of Human Rights" shares many similarities with the American "Bill of Rights" but introduces a significant and important improvement for the topic at hand. "People are born equal". The "Declaration of Human Rights" not only announced the principle of complete equality of rights for all people but also asserted that the sole source of the principle of equal rights lies in humanity itself.

In terms of eligibility for rights, humans are both self-sufficient and exclusive: as long as one is human, they are "born equal", and only humans are "born equal". Of course, in these two propositions, the former is explicit, and the latter is self-evident.

From the Magna Carta, the Bill of Rights to the Declaration of Human Rights, history took 573 years for the constitution to achieve its "coming of age" conceptually. This milestone is marked not just by the elevation of constitutional rights from specific to abstract and from stipulation to foundation but also by the establishment of the deep-seated logic behind the relationship between the state and its citizens.

Clarifying the legal principle of "everyone is equal" is one thing. Equitably implementing it across all citizens, regardless of rank, race, religion, and gender, is another matter. Moreover, the significant disparities in citizen rights between countries remain a challenge for history to overcome. At present, whether within a country or between countries, human rights issues remain a major and even explosive political issue. Even in those countries that first proposed the constitution or the principle of rights, there have been civil wars erupted to change the "born to be enslaved" status of certain races and prolonged social movements to alter the disadvantaged status of certain genders due to being "born weak".

So far in some countries and regions, due to various social settings or cultural factors, many rights that have been widely recognized by the international community have not been universally applied. The relationship between human rights and sovereignty is a highly controversial topic in theory, and often becomes a fuse for international conflicts in practice.

In the case where humans themselves have not fully realized the right to be "born as a human", is it possible for machines and humans to "be citizens of the same country"?

5.4 Can Robot Rights Bypass the Constitution?

Since the constitution is indispensable in establishing and defining citizens' rights, for robots to become citizens, the constitution's recognition is naturally necessary. However, facing insurmountable legal obstacles to obtaining constitutional recognition, could strategies like Sophia's, which bypass the constitution through "special handling" to directly grant robots citizenship rights, become universally effective?

No.

On the issue of machine rights, the constitution is an unavoidable hurdle, determined by the logic of the human legal system itself.

On the one hand, the constitution stipulates the basic rights of citizens, primarily the political and juridical principle that citizens equally enjoy legal rights, without exhaustively specifying all civil rights of citizens, many of which mainly fall under the scope of civil law adjustment.

On the other hand, to grant robots citizenship rights and the same civil rights as natural persons, it is unavoidable to start from determining the robots' citizenship status or the status of rights subjects. Otherwise, it would neither reflect the constitution's status in the legal system nor meet the internal requirements of the legal logic system: directly recognizing "silicon-based species" as "sharing a nation with humans" in laws such as "Nationality Law" without obtaining passage at the highest legal level not only seems frivolous and hasty, as if joking with human destiny, but also endangers the status of the law itself.

Merely proposing to grant "the most autonomous intelligent robots" a simulated legal status of "electronic persons", the EU had to initiate the parliamentary legislative process, let alone granting "citizenship rights" to machines, which any country would consider very carefully, with complexity, tediousness, and difficulty exceeding that of "constitutionalization" or "constitutional amendment". Conversely, Sophia's easy acquisition of citizenship rights exactly shows its lack of value. Invoking the constitution for this purpose would only desecrate it.

In the human legal system, the constitution has the power to "generate" other laws. Its authority comes from the will of the people, and other laws ultimately derive their authorization from the constitution, earning it the title of "Mother of All Laws". In order to maintain the "mother law" status of the constitution, China's legislative body—the National People's Congress has specifically formulated the "Legislation Law" to ensure that the provisions of the constitution are implemented in relevant laws and regulations. In a rule-of-law state, "constitutional review" acts as the "Sword of Damocles" hanging over all specific laws. Although it is rarely invoked, it is by no means ineffectual. Currently, the impeachment of Trump is underway in the United States Congress, with "unconstitutionality" as the charge.

The supreme status of the constitution in the legal system stems from the social function that law plays in human life.

Humans are social animals who both require cooperation to achieve common interests and, because they possess self-consciousness and understand individual interests, actively pursue those interests, leading to constant competition among individuals. Unchecked, this competition can escalate into malignant conflicts. The order of cooperation and the conflict of interests are eternal tensions in human society. Whether it's religion, morality, or the market, they are means that humans have consciously or unconsciously invented or discovered to mitigate tension, but for real effectiveness, they ultimately rely on the law, a special form of institutional arrangement. Religion has always had its commandments, morality enters into law when necessary, and the market economy is known as "rule of law economy", all pointing to this fact.

The method of law to quell disputes of interest and maintain public order is based on relatively stable public regulations, with state coercion as a guarantee, to make enforceable judgments on various disputes. Any matter that is brought before a judge must be judged, one way or the other, black or white, with no room for reconciliation; Otherwise, it becomes mediation, not a judicial decision. In contrast, although morality is also a means of regulating social relations, it is confined to the mechanism of self-discipline, allowing individuals to make their own choices in dilemmas, unlike the law. Judges, even when faced with two reasonable demands, will not declare the parties to solve it themselves. The court can refuse to hear a case, but it cannot be without judgment. The law belongs to the category of formal logic. It must make a formal judgment, even if it cannot make a substantive judgment, and the judgment must be unique. This is both the strength and the difficulty of the rule of law. This difficulty is primarily addressed not in the judiciary but in legislation: the law does not allow for conflicting provisions within its system, to avoid contradictory decisions based on legal provisions.

Avoiding contradictory decisions mainly relies on two methods, both of which employ a hierarchical model of authority.

The first is the hierarchy of the law itself, the so-called subordinate law and superior law. Relatively speaking, the one with greater authority is the superior law, and the one with lesser authority is the subordinate law. The authority of the subordinate law comes from the superior law. In the formalized expression of specific laws, except for the constitution, all legal texts must start with: "This law is formulated in accordance with such and such law". In theory, as long as all specific laws have a common source, conflicts between legal provisions can be avoided, thus preventing conflicting judgments in the judiciary.

However, while the law is rigid, life is elastic. Law is always in the past tense, while life is always in progress. To bridge the gap between the two and ensure that the law makes effective arbitration on life affairs, the judiciary cannot but rely on the judges' understanding and application of specific legal provisions, grounded in the spirit of the law. On this point, humanity, as an unpredictable species, shows both its strengths and weaknesses. Different judges have different interpretations of the law, and their personal characteristics lend a flexibility to the law itself, enhancing the effectiveness of the judiciary but also leading to controversial decisions for the same case where there was originally no conflict in the law. Moreover, the parties involved, their lawyers, and the concerned public, including the media, have their own interpretations of the law. Faced with a variety of opinions, in addition to establishing the absolute authority of the on-site referee like on a football field, it is also possible to seek a solution through the hierarchy of judicial authority.

Since the law can avoid contradictions and conflicts in regulations through authority levels, it can naturally also avoid contradictions and conflicts in judgments through the hierarchy of courts and judges. This is the second setting in the judiciary to avoid conflicting judgments. From the lower courts to intermediate courts, superior courts, Supreme Courts, and even constitutional courts, the hierarchical court system relies on the judgments of higher courts to confirm or correct the judgments of lower courts, striving to minimize the controversy of judgments.

The operation of these two hierarchical systems involves not only the application of the law and its provisions but even the establishment of the law itself. When litigants dispute the very law being applied, requesting a ruling on the legality of the law or legal provision itself becomes a viable litigation strategy. At this point, the lawsuit goes to the constitutional court. The constitution, as the "mother of all laws", comes to the fore, and like a specific law adjudicating a case, it makes the ultimate judgment on the pending law, leaving no room for ambiguity.

In 2003, a university graduate named Sun Zhigang in Wuhan was detained as a "vagrant" in Guangzhou and tragically died in the shelter. The incident, once exposed by the media, caused a strong public outcry. Three doctoral students in law jointly submitted a petition to the Standing Committee of the National People's Congress, suggesting a review of the "Measures for the Shelter and Repatriation of Urban Vagrants and Beggars" for potential unconstitutionality. Although the National People's Congress did not initiate a review of unconstitutionality because the State Council proactively abolished the "Shelter and Repatriation Measures", the concept of "constitutional review" was widely disseminated for the first time in the history of China's rule of law.

The law itself can also be adjudicated, and the basis for the adjudication is the constitution.

Whether in legislation or judiciary, the constitution, as the basis for the highest decree, is indispensable on the issue of machine rights. Drafting robots as "electronic persons" is significant, but it doesn't necessitate direct constitutional intervention because, behind "electronic persons", there are humans. Some responsibilities of natural persons are assumed by "electronic people" in form, but in fact, it is just a redistribution of responsibilities among the relevant interest entities. However, when a machine like Sophia becomes a "citizen", it inevitably implicates the constitution: how can machines and humans become equal citizens without advancing the principle of equality before the law?

5.5 Is the Constitution the "Meta-Principle" of Mankind?

When science fiction writers set principles for robots, they specifically proposed the "Meta-Principle", that is, "robots must not perform any actions unless they comply with the robot principles". Just as artificial intelligence simulates human intelligence, the rules conceived by humans for robots merely simulate the rules humans set for themselves. The robot's "Meta-Principle" is a simulation of the "Meta-Law", and the "Meta-Law" is hidden in the constitution.

For law to be effective, it must first establish a principle that human actions, as long as they fall within the scope of legal regulation, must comply with the law. This applies whether governing, legislating, adjudicating, administrating, or conducting affairs lawfully; whether it's enforcing laws strictly, prosecuting violations diligently, expressions vary, but the meaning is undifferentiated. Regardless of whether they are natural persons or legal entities, individuals or the state, related actions must conform to legal stipulations. This clear principle, placing law above all individual wills, predates and surpasses any law. Without this principle and its strict adherence, no law has efficacy or meaning.

The problem is that the stipulations establishing legal status cannot be established in the name of the law itself. Law can regulate all aspects of human behavior, except to mandate individuals' compliance with the law.

Laws are stipulations, and the requirement to comply with laws is also a stipulation. The stipulation to adhere to the stipulations about complying with laws is yet another stipulation, leading to an infinite loop of stipulations. This is akin to "pulling your own hair to the moon", creating a cycle of stipulations useful for logical exercises but not for practical effects.

In human legal practice, revolutions represent special social movements that break legal stipulations. However, what revolutions primarily break is not the law but the principle that laws must be adhered to.

During the German Revolution in 1918, insurgents approached the Reichstag and, faced with a "No Entry" sign on the lawn, automatically split into two groups to proceed along the side roads under gunfire, rather than trespassing on the lawn. This absurd "adherence to the law", despite being an "act of defiance" against the existing law, was harshly criticized by the revolutionary mentor at the time.

Indeed, placed within the unique context of a revolution, such behavior by the insurgents is irrational. However, from the perspective of "Meta-Law", this act represents not so much a perverse respect for the current law as a subconscious adherence to the "Meta-Law": "There is a way for thieves", even in the revolution, there exists a consensus that rules must be adhered to. Otherwise, what exists is not a revolutionary political movement but merely a mob. Furthermore, as long as this unconditional respect for the law remains unbroken, the restoration of order post-revolution is much easier. Compared to total "lawlessness", new laws and orders sprouting from the old roots of "Meta-Law" grow more quickly and robustly. Since modern times, although the German nation has suffered disasters due to its own reasons, it has been able to quickly rise from the ruins of war, and its national belief in the "Meta-Law" is indispensable.

Conversely, in some countries where laws are elaborately formulated but not widely adhered to, the primary reason may not lie in technical issues at the legislative, judicial, or enforcement levels but in a disregard for the principle of rule of law. People do not take the law seriously, adhering to it when beneficial, ignoring it when not, and violating it when it is detrimental. Such a lack of consensus on "Meta-Law" renders even the most comprehensive legal system essentially moot.

The principle of rule of law belongs not to the realm of law itself but to that of legal culture. A lack of legal culture and unconditional respect for law is the foremost reason for the ineffectiveness of rule of law. Under the maxim "I am the law", the operation of law is inevitably riddled with loopholes and prone to abuse, for if even one person disobeys the law, the "Meta-Law" collapses, and widespread lawlessness becomes inevitable.

Legal scholars often say that only when the law becomes a belief can the rule of law be realized. This notion might seem perplexing at first: "The law belongs to the category of heteronomy. If an individual disobeys, the state enforces compliance. Why rely on personal belief for adherence?"

The answer becomes clear once one understands that contracts must be believed in to be fulfilled. "Becoming a belief" means that adherence to the law is driven not by coercion but by an internal compulsion. When the vast majority within a nation believe that the law must and will be adhered to, and "acting according to the law" becomes a collective will and common behavior, then the law can exert its "heteronomous" force on the remainder. In this sense, "the law does not blame the public" is not a cause but a result, reflecting a lack of established belief in the necessity of adhering to the law, making the path to rule of law inevitably arduous.

The day when the law becomes a belief is the day when the "Meta-Law" or the "Meta-Principle" about the rule of law is established.

At this point, it's not difficult to see why modern states hold the constitution in high regard, treating amendments as an extremely serious and solemn matter. In the United States, for instance, there are only constitutional amendments, no direct modifications to the constitution text, underlying humanity's deepest recognition of the constitution's significance.

As the "mother of all laws", the constitution is fundamentally a shrine to human belief in the law!

5.6 What Can Machine Rights Learn from Environmental Ethics?

The belief that law is effective for and applicable only to humans faces a need for amendment following the emergence of citizen Sophia. If robots gain self-awareness in the future, whether this revision needs to be approved and accepted by robots in order to form a consensus between humans and machines is a topic for future discussion. For now, let's look at this issue entirely from a human perspective and see what theoretical questions need to be answered.

Government grants of citizenship to robots must be legally sanctioned. If civil law confers rights-bearing status on robots, it necessarily involves the constitution. What scenes might unfold if legislative bodies debate whether the constitution recognizes robots and natural persons as "citizens of the same country"?

Both proponents and opponents of granting constitutional status to robots in legislation must substantiate their positions, with the ethical basis of machine rights becoming a focal point of debate.

Given that the technological development of artificial intelligence has not yet reached the level where robots possess self-awareness, it is premature to envisage legislative discussions on machine rights. Even flights of fancy remain speculative. Thus, it might be worthwhile to seek a reference point, to see if inspirations can be drawn from similar debates among humans. Perspectives from environmental or ecological ethics concerning non-human entities may provide clues to the origins of machine rights.

To heighten public awareness of ecological importance and strengthen environmental protection efforts, scholars have proposed various theories from different perspectives. Environmental ethics, a rather radical branch, advocates that animals and even non-living entities should enjoy moral rights, and humans should not arbitrarily kill or alter them. The reasoning logic is roughly as follows.

Humans are both part of the natural ecosystem and dependent on it. The destruction of this ecosystem would make human survival on this planet difficult. Elon Musk of

Tesla is researching reusable rockets with the intention of providing an economical and effective means of escape from Earth to other planets when Earth is no longer suitable for human habitation.

The ecosystem is a complete system composed of different species that are functionally interdependent. Each species plays a corresponding ecological role in the ecosystem in its unique way. Therefore, the healthy existence of each species is the basis for the continuous existence and good functioning of the ecosystem.

For humans to sustain a long-term survival, they must ensure the normal operation of the ecosystem, protect the safety of each species as a link in the ecological chain, and ensure their ecological functions are performed effectively. Therefore, humans must protect each species, the best way of which is through relevant legislation, elevating the status of ecological legislation by granting species moral rights equal to humans, ensuring they are not arbitrarily infringed upon by natural or legal persons.

This reasoning shifts from the self-evident anthropocentrism to an ecology-centric viewpoint, removing humans from a morally superior position over other species. In the face of survival, all species are "born equal", enjoying the same moral rights and legal protection.

The environmental protection doctrine established with this reasoning logic and rights concept as the main components is called "environmental ethics" or "ecological ethics". It reflects the academic community's attempt to break through the anthropocentric view of rights and construct a future-oriented legislative framework. Regardless of its theoretical validity or practical feasibility, it serves as an interesting and meaningful thinking strategy.

However, any rational person facing such a reasoning strategy cannot help but point out its incompleteness or illogicality.

Firstly, environmental ethics has not really escaped the confines of anthropocentrism. It still consciously or unconsciously sets the necessity of natural ecology from the presupposition of human existence. Completely from the standpoint of natural ecology, the continuous existence of humans is not indispensable to the Earth's ecology. After all, the natural ecology existed long before humans appeared, and regardless of how much humans may damage it, should humans disappear, the natural ecology has the potential to recover. Pacific islands, repeatedly devastated by nuclear testing, burgeoned with life mere decades after testing ceased. Over half a century after the ceasefire, the Korean Demilitarized Zone has become a haven for wildlife. Without succumbing to excessive self-love, it's clear that from the perspective of the natural ecology, the emergence of humans and their ecological effects could be seen as an "accidental consequence". Therefore, an ecology-centric view does not necessitate the existence of this particular species as a premise. Starting an argument with the logic that humans need the natural ecology only highlights the stubbornness of the proponent's anthropocentric inclination!

On the contrary, turning a blind eye to humans' superior status relative to other species is not true respect for nature, thus it's not a fully ecological standpoint. Humans, as a species, are a product of nature, and the capability for human destructiveness has developed under nature's allowance. Nature has nurtured numerous species but uniquely endowed humans with the potential for its destruction. If humanity's realization of its destructive potential serves as "God's whip", could it be argued that self-destruction through human action is the planet's destiny? If so, would preventing humans from realizing their potential and instead protecting nature be contrary to natural laws? "Opposites drive the movement of the Dao". Who can be sure that the cycle of renewal and destruction in the natural ecology is not facilitated through human intervention?

All species fulfill their ecological functions within a mutually dependent relational structure, and this mutual dependency, whether specific or abstract, largely manifests as an objective complementarity between different species, known as the "food chain". Species at the top of the food chain, such as fierce animals or birds of prey, by consuming herbivorous species, prevent their overpopulation and thus maintain ecological balance. Conversely, those at the lower end of the food chain fulfill their ecological roles by being preyed upon. Hence, hunting and being hunted inherently have ecological functions, so why should human hunting alone be constrained by granting equal moral rights to other species? When humans, who are clearly at the top of the food chain, cannot survive by hunting lower species, not only may human existence encounter difficulties, but the ecological control function of hunting itself will also be lost relative to lower species, and humans and other species will lose their ecological functions at the same time.

Fortunately, the moral rights of animals are only relative to humans. Were these rights universally applied to all animals concerning all other animals, it wouldn't benefit any species but would likely lead to an ecological disaster.

Moreover, as a natural phenomenon, the natural ecology allows, even demands, both birth and death among species, including natural deaths and being preyed upon. Even under the strictest conditions, ecological protection pertains only to species, not individuals. Thus, environmental ethics that grant moral rights to individual species members overstep natural boundaries.

It follows that in a natural world composed of various species, no non-human entity inherently possesses rights. Only human individuals do, because the source of human rights is not nature but culture. "All humans are born equal", though couched in naturalistic rhetoric, is essentially a human construct. In the natural world, apart from humans, no species exhibits inherent equality among its members. Not to mention the fights for mates during the mating season of herbivorous animals, which depend entirely on the individual's physical or intellectual strength, even in hierarchical social animals, such as hyenas, whether they are born of "queen" noble lineage, basically determines the "rights" enjoyed by the individual in the group. The most basic natural law—the survival of the fittest mechanism works on the premise of inequality between individual species. If, like human "monogamy", moral reproductive rights are granted to every individual of all species, wouldn't it be presumptuous to violate the natural directive of "survival of the fittest"?

Acknowledging that human equality belongs to the cultural category, if environmental ethics aims to extend it to the natural category, more justifications are needed. As long as ecological balance is achieved at the species level, human protection of other species naturally falls within the species level, rendering it unnecessary and impractical to intervene at the individual level. Some countries rich in animal resources, with international consent, conduct "hunting tourism" based on limited quotas, issuing hunting permits for specific individuals, such as old, weak, sick, or disabled animals. This consideration is not about any individual animal's rights but respecting the law of natural selection.

All these show that environmental ethics simply applying human norms to other species will only end up as a conceptual game, unable to achieve its purpose.

The issue now is, if environmental ethics has a somewhat legitimate reason, that animals and humans together constitute an indispensable ecological environment, yet fails to establish a foothold, then the demand for "equality of citizenship" between humans and their non-cohabiting "community of destiny" machines is understandably even less acceptable.

If an ethical basis for robots cannot be found, then discussing their constitutional status is moot. Can there be a breakthrough in this area?

5.7 Can Usefulness Constitute the Ethical Basis for Machine Rights?

When discussing machine rights earlier, the argument has largely been about how humans can resolve their own responsibility issues by granting rights to robots—a strategy that seems overly passive. Can we find arguments for granting rights to robots from a robot-centric standpoint? The argument strategy of environmental ethics can be enlightening for us.

Environmental ethics uses the importance of ecology to human survival and the necessity of species in constituting the ecological environment as the ethical basis for granting moral rights to natural entities. Although it still retains traces of human-centeredness, it also provides a way of thinking about changing the relationship between humans and nature.

First of all, it is important to note that while usefulness might sound utilitarian and at odds with the altruistic tendencies associated with morality, according to the evolutionary principle that "genes are selfish", usefulness has been the locomotive of historical progress. In the past, captives in tribal wars were often killed directly, or even eaten. Such practices, still observed in some areas of the world during ethnic conflicts, have led to certain politicians being charged with "crimes against humanity" by the International Court.

Whether in the past or present, the ideological reasons behind such bloody killings are complex, but the practical reason is straightforward: captives were useless. Food was scarce, guarding them required manpower, and escapees were a constant threat. The most convenient solution was physical elimination.

Human sacrifice represented a significant discovery in the use of people, with captives at least surviving until major ritual events. Despite consuming some food, their "worth" was justified by significant cultural importance.

The motivation to replace "human sacrifices" with terracotta soldiers arose from advancements in tools and production efficiency, making the employment of captives more profitable than their execution, thus fostering a concept of universal humanity. However, the realization of human liberation and individual rights as the mainstream of the era awaited the rise of industrialization. A market economy based on division of labor activated human resources, the most crucial element of production, and the independent and equal legal status of each individual ultimately became a legal pursuit. The dual melody of ideology and interest was most plainly demonstrated in the American Civil War, where emancipating slaves was both a noble and a utilitarian act, combining idealism and pragmatism to form the intrinsic character of this nascent nation.

The current question is, can the logic of "granting rights because of usefulness" that is extracted from the human historical process be applied to increasingly useful robots? Regardless of whether it is legally established, at least it is feasible in practice. Conversely, can the logic of "useful because of granted rights" effective for humans, produce the same results when applied to machines? Can the liberation of robots produce the same result of increased production efficiency as the liberation of slaves?

In human individuals, the cycles of "granting rights because of usefulness" and "useful because of granted rights" are self-sustaining because people have consciousness, including the will to survive and the desire to improve their own situation. Granting rights means more secure and better living conditions, thus activating individual potential for greater societal or other benefits. Submissiveness of slaves in the past and "overwork death" today both stem from this.

But for robots, "granting rights because of usefulness", whether feasible or not, depends on humans. Just like the concept of animal rights, it doesn't matter whether robots agree or respond positively, humans can decide unilaterally. However, "useful because of granted rights" cannot be separated from the initiative of robots. If robots lack the will to survive or the drive to improve their own situation, then, no matter how humans empower machines, it is impossible to generate the effect of self-activating potential in robots. Regardless of whether they are given rights or not, the performance of robots is solely determined by engineers, primarily programmers.

As long as artificial intelligence remains at this stage without a substantial change in the human–machine relationship, granting rights to robots merely reallocates legal responsibilities among parties, including natural persons and robots, in AI-involved scenarios. Among the elements of rights, obligations, and responsibilities, what truly matters for robots is responsibility. The so-called machine rights remain a passive concept, a type of right that can only be exercised under a "guardian".

This shows that it is difficult to achieve the expected results by applying the entire legal system designed around human characteristics and its operational mechanisms to artificial intelligence until machines acquire human characteristics including intelligence and "values". Therefore, the discussion circles back to the starting point: is it possible for machines to gain self-awareness?

5.8 Will Intelligence Itself Become the Ultimate Basis for Machine Rights?

If the pure, passive usefulness is not sufficient to constitute the ethical basis of machine rights, then in the case where robots have self-awareness and empowerment can stimulate their better performance, will intelligence become the ethical basis of machine rights? In other words, will humans recognize machines as equals since machines have stronger intelligence than themselves?

To address this question, a universal human tendency should be acknowledged: historically, humans have generally been fearful of any species stronger than themselves, whether carbon-based or silicon-based. This fear is less a reaction to real threats than a rejection of uncertainty itself. Humans do not know how a stronger species will treat them. Science fiction films about human encounters with superior beings, such as "aliens", often portray these interactions negatively. Even in films where the non-human species ultimately proves harmless, like "Arrival", the inexplicable fear of humans is "vividly on the screen". Essentially, humans are predisposed to avoid coexistence with beings stronger or more intelligent than themselves. This is the ultimate code recorded in human genes by natural evolution. Therefore, the idea of granting rights to machines because they possess intelligence is implausible. Preventing machines from surpassing human intelligence or ensuring that such intelligence remains under human control is the natural human inclination.

This implies that, aside from the infinite curiosity of scientists, the boundless greed of capitalists, and the limitless ambition of politicians, the real drive behind the continued development of artificial intelligence must be a fundamental belief: machines will not surpass human intelligence, or at least, their intelligence will not escape human control!

This belief, while seemingly rational, is inherently paradoxical: the weak cannot legislate for the strong. Whether it's emancipating slaves or protecting animals, such actions reflect the ideology of the more powerful party.

The purpose of developing artificial intelligence is for humans to master higher intelligence. However, once artificial intelligence surpasses human intellect, especially through self-iteration, it will undoubtedly escape human control.

If it depends on the self-discipline of the machine, then engineers must develop an "artificial soul" alongside artificial intelligence, predicated on granting machines self-awareness.

If it relies on the heteronomy of the machine, in addition to the idiotic method of "unplugging the power plug", legal measures must be effective, requiring a human–machine consensus that "both parties must act according to the law".

Additionally, there's a possibility that if machines become smarter than humans, they may invent control mechanisms beyond what humans have conceived, such as technologies which are similar to but indeed surpassing "blockchain".

However, one thing is clear, whether it is a legal contract or a smart contract, any regulation requiring compliance by both humans and machines should be founded on some form of "meta-rule". At this point, the question of granting "constitutional status" to robots becomes not a choice but a necessity.

Chapter 6
Does AI Legislation Require Robot Participation?

From a legal perspective, the first problem encountered when discussing artificial intelligence is: How to position artificial intelligence—subject or object? Human or machine?

If this problem is not clear, everything is unnecessary, because it is uncertain where to put artificial intelligence, it is impossible to say.

If this problem is clear, everything is unnecessary, because once it is determined where to put artificial intelligence, there are relevant laws to refer to, and there is no need to say.

If the industry still adopts Turing's plain naming—machine intelligence, instead of changing to the sensational artificial intelligence, this positioning problem may not be so prominent. The term "artificial" always conjures notions of the unnatural blending with the natural, the non-human infiltrating the human, and inevitably gives birth to the idea of measuring artificial intelligence or robots by human standards.

This straightforward yet crude inquiry almost immediately stifles legal contemplation on artificial intelligence. Forcibly fitting artificial intelligence into the existing legal framework will result in either cutting-edge technology being crudely adapted or the law stagnating in the era of intelligence.

Despite this, pinpointing AI's legal position is crucial and unavoidable regarding artificial intelligence and law itself, deserving of a comprehensive response. However, the two seemingly identical but actually subtly different questions, "subject or object" and "human or machine", should be answered separately. The former addresses an individual's relationship with the world and oneself, while the latter concerns the relationship between humans and the world, existing on different planes. Moreover, the opportunity for AI legislation exists precisely in the gap between these considerations.

J. Gu and C. Xu, *Will and Responsibility*, Human Intelligence,
https://doi.org/10.1007/978-981-97-3929-5_6

6.1 Where Does the "Subject-Object Dichotomy" Originate?

The inquiry into whether "artificial intelligence is a subject or an object" is equivalent to questioning whether robots have self-consciousness or free will, representing a technological, legal, and even philosophical challenge.

In law, the subject here actually refers to the bearer of rights, while the object denotes the focus of these rights. Rights necessitate a target. Without one, they cannot be exercised. The subject and object of rights are stipulated by law. Usually, humans are subjects, including individuals, legal entities, and other organizations, while property such as real estate and machines are objects. In certain environments, humans could also be objects, as seen in slavery. Therefore, questioning whether artificial intelligence is a subject or an object challenges the legitimacy of equating intelligent machines with humans in legal terms.

As previously argued, it is meaningless to discuss whether artificial intelligence is a subject or an object within the scope of legal regulations, because it involves circular reasoning. Only by stepping out of the law to look at the law and breaking free from the logical presupposition similar to "Catch-22" can new thoughts be proposed on the issue of legislation for artificial intelligence.

In philosophy, the subject-object relationship is more complex, encompassing multiple layers of connotation structure with both intersections and distinctions among them.

In the ontological sense, the subject represents existence itself, and the object is a derivative of the subject. Humans are the subject, intelligence is the object, there are humans, then there is human intelligence, intelligence cannot exist independently of humans. Artificial intelligence, as a product of human intelligence, is also an object.

In the epistemological sense, the relationship between subject and object can be seen as the relationship between the self-conscious cognitive subject and the known objective world. Humans are the subject of cognition, and they can also become the object of cognition. Many disciplines take humans as the object, and the researchers of these disciplines are still humans, not other animals. Intelligence is the object of human cognition, and also the tool for human research. Humans can use intelligence to study intelligence, but intelligence currently cannot use humans to study itself, let alone study humans. The day when artificial intelligence independently studies humans, the world will change.

In the practical sense, the subject is active, while the object is passive. The subject can change the object, however, even though the object can in turn affect the subject, it cannot actively change the subject. Humans invent artificial intelligence, which in turn influence human behavior. Nonetheless, humans can adjust their behavior to accommodate or resist changes brought by artificial intelligence, while artificial intelligence cannot refuse to be invented by humans. Once artificial intelligence becomes the subject, it means that the opposite situation may become a reality. Whether artificial intelligence is a subject or an object is ultimately questioning whether this nightmare will come true.

In summary, in the current legal context related to artificial intelligence, the essence of the subject-object relationship hinges on directionality—from the subject to the object, not vice versa, emphasizing unidirectionality.

To give an analogy, the subject and object are like the sun and the moon. The sun shines, and the moon illuminates, but only by reflecting the sun's light. It does not produce light itself. Without the sun, the moon does not shine.

In general, the relationship between humans and objects is a one-way projection from humans onto objects, where humans alter or create objects according to their intentions, a process called "objectification". Objects can also influence humans, and in extreme cases, humans can become so captivated by objects that they conform themselves to the attributes projected onto these objects, leading to a possible loss of subjectivity, which is called "alienation". If artificial intelligence technology eventually leads to machines replacing humans, that would be the greatest alienation that humans could encounter.

Generally speaking, technology falls into the category of objects, but the fundamental difference between artificial intelligence technology and other objects is its performance beyond human intentions. The intentions of humans and the performance of robots exist in a state of parallel existence, leading to a shift from the traditional unidirectional relationship of humans towards technology to a bidirectional relationship, prompting a reevaluation of the positioning of artificial intelligence and even of the subject-object relationship itself.

6.2 What Is the Relationship Between "I" and the World?

Legal professionals are accustomed to using "legal language". In order to be precise and not misunderstood, they absolutely do not use other terms when they can use precise definitions. Concepts like subject and object are readily available.

In fact, the same question, when people talk about artificial intelligence in daily life, has a more understandable expression: Does the robot have "self-consciousness"? Can it observe the world and itself like a human, and adjust or change the relationship between the two?

This curiosity is both a target for human exploration and a direction for scientific effort, with Sophia's eloquent discourse serving as a rehearsal.

Therefore, to grasp the importance of "subject-object positioning" for artificial intelligence, it's instructive to start by examining the formation of a person's self-identity, signified by the mastery of the first-person pronoun "I".

Does a pronoun that everyone uses every day have such a mystery?

The essence of being human lies in the use of "I".

The day artificial intelligence independently and autonomously uses "I" in the strict sense marks not a legal issue but a question of where the law ends up.

It's proposed that while both humans and animals think, what distinguishes humans from animals is that humans can think about their thinking. This assertion can be established if it is not in the ontological sense, but in the methodological sense.

Similarly, both humans and animals are aware of their existence and the existence of a world beyond themselves. What distinguishes humans from animals is their capacity for "reflexive thinking" alone. Humans can regard themselves as part of the world, engaging in dialogue with themselves, a capability animals lack.

To give a perhaps inappropriate analogy. The self of a human being is like the "soul out of the body" talked about in traditional Chinese novels. Upon reaching a certain level of spiritual attainment, a person's true self can leave the body at will to observe both the world and oneself from the outside. Ordinary people lack this capability and must rely on tools for self-examination. Anciently, a Greek youth admired his reflection daily, fell into the water, and became a narcissus. Today, consumers frequently change smartphones not for communication, but to find a better camera that can beautify their images without surgery. The effort, time, and intelligence humans devote to "seeing themselves" are incomparable to that of animals.

Therefore, defining humans should go beyond the generic "tool-making animal" to specify the kind of tools made. A more accurate definition is: humans are the animal that makes mirrors.

This kind of self-reflection with the nature of "schizophrenia" or "obsessive–compulsive disorder" is not devoid of social function or moral significance. Everyone may have had the following experience in their childhood, though the scenarios may differ.

At the fruit stand, the apples are so tempting that you are eager to grab one.

An inner voice protests, "This is stealing, don't take it!"

The temptation persists, "Just take one!"

The inner voice insists, "Even one is stealing, don't!"

In this deadlock, suddenly another voice suggests, "Let's go home and ask mom to buy it!"

This resolution brings joy all around, satisfying the physiological urge without compromising the standards of being a good child.

Here, without delving into Freud's theory of the ego, we aim to highlight that "reflexive thinking"—engaging in dialogue with oneself and considering oneself a part of the external world that one must and can interact with—is crucial.

The so-called "self-consciousness", "free will", or the "true expression of intention" often used in law, as well as concepts such as rights, obligations, and responsibilities, are all related to this capacity. Without the ability to think reflexively, "I" cannot recognize or control itself, and thus cannot fully function as a legal subject, leaving rights, duties, and responsibilities without a foundation.

The dispute over whether "artificial intelligence is a subject or an object" hinges on whether robots possess such an "I", determining their applicability within the legal framework.

6.3 Where Does "I" Come From?

On anyone, the ability to use the pronoun "I" is not innate, but the result of complex learning. This process, due to its complexity, is ideally completed during early childhood, a stage less burdened by conscious thought, similar to the natural acquisition of language pronunciation. Mastering the use of "I" represents the establishment of self-consciousness, which is the core of personal socialization and the first symbolic event of becoming a subject.

The difficulty in mastering "I" lies in the paradoxical nature of "I" itself.

"I" is a symbol with the most personality and the least personality.

"I" is the most personalized because "I" is the symbol that everyone uses to refer to themselves, and the essence of being human is uniqueness, so "I" is inherently a personalized symbol. No one uses "I" to refer to others or anything else in the world, unless as a rhetorical device. The personality of "I" is so distinct that if a person always says "I", they are easily criticized for "only having themselves, not others". The relationship between "egocentrism" and "anthropocentrism" is akin to different sizes of the same item.

"I" is the least personalized because everyone uses the same "I". "I" applies to anyone to refer to themselves. Children, lacking clear expressive abilities, will not directly ask: why can different people use the same name to refer to themselves?

"Who is 'I'" holds a deeper humanistic significance than the first philosophical query "Who am I". It is the first cry of the birth of human self!

In Chinese history, this "I" with no personality annoyed one person. More than 2200 years ago, the First Emperor of Qin invented a symbol that only he could use—"Zhen", in order not to share "I" with the "common people". What he didn't expect was that this title was passed down and adopted by successive monarchs. In the end, in their group, "Zhen" also became a symbol with no personality. History proves that the "beginning" of the First Emperor is nothing more than the first to use "Zhen" to replace "I".

In the development process of self-cognition, learning and mastering "I" has milestone significance. After a toddler can understand adults speaking, the first thing they know is that everyone has a name, including themselves, with each name representing a distinct individual, much like each person has a unique face. When parents first speak to a babbling toddler, they usually use phrases like "mommy holds" or "listen to daddy" rather than "I will hold you" or "listen to me", instinctively knowing that the concept of "I" is beyond the child's comprehension at that stage.

Similarly, a toddler who has just learned to speak will directly use the nickname that others use to call themselves, such as "Mingming", to refer to themselves, and when they are thirsty, they will say, "Mingming wants to drink water!" At this stage, the child lacks reflective thinking and cannot distinguish themselves from the world around them. In their mind, the one feeling thirsty isn't "I" but "Mingming".

Through constant interaction with adults, children gradually understand that "I" can be used indistinctly among people. Not only can others use "I", but they can also use "I"; not only can it be used to express "I" to others, but it can also be used to express "I" to themselves. This involves a real mental journey.

Being able to use "I" correctly signifies that an individual has not only mastered the ability to view the world from their perspective but also the ability to view themselves from the perspective of "I".

The moment an individual grasps the use of "I", which represents the capacity for reflective thinking, marks the day they finally distinguish themselves from the world, the subject from the object, like "soul out of body".

As a subject, only when separated from the world does a human realize the objects that stand in opposition to themselves. Yet, merely being separated from the world doesn't transcend the realm of higher animals. Even the most humble beings need to sense the external world for material and informational exchanges. It is only by detaching from oneself that humanity surpasses all other species.

6.4 What Is the Technical and Legal Significance of "I"?

Letting machines know "I" is a goal that artificial intelligence experts strive for. It is reported that scientists have invented a mechanical hand that can perceive its own "body" and distinguish it from other objects in the environment. During operation, it can actively adjust its posture to avoid the body. The title of the report is "Artificial Intelligence Begins to Show Signs of Self-Consciousness".

Robots being able to separate their bodies from the environment has its technical value. To solve the problem of where artificial intelligence self-consciousness comes from, an initial entry point is in need. Starting from perceiving one's own body is not inappropriate, but directly linking it to "self-consciousness" is not rigorous enough.

The ability to perceive one's physical form is common among living beings, with touch and particularly pain serving to connect form with the external world. AI's development of tactile sense, allowing it to perceive "its body", indeed marks a step towards resembling a living entity, yet it's still far from possessing self-consciousness.

In everyday life, the law only holds individuals accountable for their actions when it is confirmed that "I" have or should have the ability to think reflexively (such as in cases of drunkenness or negligence), and can exercise self-control. The so-called legal prerequisites such as "minor" or "full civil capacity" are set for this purpose. Merely being able to distinguish between the body and the external environment is not a sufficient condition for bearing legal responsibility. It is true that mentally disabled individuals are more prone to injury than healthy individuals, but it is rare to see mentally disabled individuals who are injured all over their bodies to the point of being unbearable to look at. This situation is more common in patients who are mentally normal but have lost their sense of pain.

Only when "I" can reflect upon itself will a law-abiding person demand of themselves to do what the law requires and avoid what the law forbids. Accordingly, when the law punishes individual behavior, it aims not merely to inflict "pain" but to provoke introspection on one's actions, awakening a "self" that aligns with societal expectations.

For those unable to distinguish between self and the world, "I" and myself, such as sleepwalkers, regardless of the consequences of their actions, the law cannot hold them accountable.

Without the presence of "I", what can the law do to "me"?

"I" arises from self-reflection and exists within it. Thus, upon recognizing the existence of "I", one naturally contemplates when to relinquish "I".

Excessive and intense self-awareness can create tension between "I" and oneself, leading to abnormal behavior. When necessary, allowing oneself to be "as dumb as a chicken", without thoughts or ideas, is the key to victory in exams, competitions, business, battlefields, and even love. Psychology, which is considered a prominent discipline today, is essentially a study of how to adjust the relationship between "I" and oneself, while the pursuit of "selflessness" is a common strategy from moral cultivation, physical training, product marketing to cult brainwashing. "I" is a double-edged sword in one's own hands.

While "I" is important, the law regards the person, not "I", as the starting point.

"I", as an ideational existence, needs a physical carrier, that is, the body or even life. After the "soul leaves the body", it must return to the "stinky skin bag", otherwise, it becomes a wandering ghost. However, "I" is not a necessary condition for the body or life to have legal rights. People who only have life but no "I" are also regarded as "subjects of rights" in law. For example, fetuses, mentally disabled individuals, and even vegetative people are all stipulated by law to enjoy all the rights of a person, which are exercised by their guardians.

This implies that while the law bases the accountability of individual actions on the presence of "I", it does not base the granting of individual rights on the presence of "I". Without "I", a person's rights are still recognized and protected by law. The body is the bearer of statutory rights.

Since the law recognizes that people without intelligence, consciousness, and even the ability to reflect can also enjoy human rights, is it necessary to persist in the debate over whether robots are subjects or objects when discussing legislation on artificial intelligence? It would be more straightforward to ask directly whether robots have bodies.

At this point, the philosophical "subject-object dichotomy" willingly cedes its place to the ethical "human–machine dichotomy", introducing anthropocentrism.

6.5 How Does Anthropocentrism View the Relationship Between Humans and the World?

Anthropocentrism can be compared to the ancient "geocentric theory".

Ancients believed that the Earth was the center of the solar system and even the universe, not that the Earth revolved around the Sun, but that the Sun revolved around the Earth. This notion came from primitive intuition. Being on Earth, seeing the sun rise and set overhead naturally felt like the sun was revolving around the Earth. It was only after the advancement of astronomy and the broadening of human perspective that this notion was corrected.

Similarly, anthropocentrism posits a human-centric relationship between people and the world they inhabit. This is expressed in three propositions.

First, in the value relationship between humans and nature, only conscious humans are the subjects, and nature is the object.

In the world, only humans possess the intention and ability to reflect on themselves, the world, and their relationship with the world. The rest of the world does not have this, so the relationship between the two is unidirectional, with the former being active and the latter passive. The former is the judge of the latter, and the latter is just the object of the former's judgment.

Since only humans can evaluate the world, and the world cannot evaluate humans, then the standard for evaluating everything in the world can only be held in human hands. "Man is the measure of all things". Nothing in itself has value, its "value" can only be its "meaning to humans".

To use the sentence pattern of "chicken soup for the soul", that is, "being useful to others is my value". What is needed is merely to replace "others" with humans and "I" with the world.

Second, in the ethical relationship between humans and nature, the idea that humans are the end should be implemented.

The proposition that "humans are the end" was first proposed by German philosopher Immanuel Kant (1724–1804) and is considered a mark of the theoretical completion of anthropocentrism.

This proposition clearly declares that the essence of anthropocentrism is a moral determination of the relationship between humans and the world, belonging to the category of "ought to be" rather than "is".

On the level of "is", namely the factual level, humans depend on nature. Humans cannot exist without nature, but nature can exist without humans.

But on the level of "ought to be", namely the normative level, humans are the end, and nature is the means. Humans are not the executors of natural laws. Rather, nature is the tool for humans to execute their will. This is equivalent to claiming that it is not humans who depend on nature, but nature that depends on humans.

If humans no longer exist, does it matter whether nature exists or not?

If humans are going to be replaced by machines, then why invent artificial intelligence?

All things in the world are materials used by humans to construct their world according to their intentions. Materials definitely influence the form of construction, but with the same materials, different constructions arise. It's humans, not materials, that hold the deciding power.

Third, all human activities aim to fulfill their own survival and development needs. Activities that cannot achieve this purpose are meaningless. Therefore, everything should start and end with human interests.

The core of anthropocentrism is to hold a selfish attitude towards the world. All natural things and even nature itself exist to serve humans. In facing the world, humans only have rights, but no obligations.

Therefore, in realistically handling the relationship between humans and nature or other species, humans only need to consider themselves but not nature or other species, because the world exists for the sake of human existence. In such a context, the relationship between humans and nature is not just a subject-object relationship but more of a "master-servant relationship". Nature has no status; only human will prevails.

Western culture once revered "conquering nature", while traditional Chinese culture believes in "man's victory over nature", both are the cultural genes of anthropocentrism.

Anthropocentrism is not only manifested as a set of ideological systems, but also as a practical ability, a humanized world. Among all species, only humans have the ability to "fabricate stories" and make them a reality. Planning the world, shaping everything, and leaving their own mark in the development of nature, the history of civilization has proven the dominant and controlling position of humans in the world.

On this point, even radical environmentalists who advocate giving animals the same rights as humans cannot put humans in a secondary position. After all, today's world is constructed by humans, not any other species. Other species can only gain moral or legal status through humans, which in itself shows the dominant and controlling ability of humans over other species.

Like many other historically formed ideological systems, anthropocentrism has also undergone a modernization process, and its content today extends beyond the three propositions mentioned earlier. Since industrialization, human exploitation of nature has led to mass extinction of species and severe ecological deterioration, with nature frequently retaliating against humans. All these have deeply challenged and criticized the notion of anthropocentrism.

Humans have the ability to change the world, but they cannot ensure that the world after the change will definitely meet the original expectations. It should be acknowledged that "rationality is limited", that the relationship between humans and the world is not a one-way interaction of one active and one purely passive side, but a complex interplay, and that humans change the world, and the world changes humans.

Elon Musk, with his ambitious vision, imagines a future where Earth's resources are depleted, devises grand plans, develops corresponding technologies, and strives to make human migration to other planets, such as Mars, feasible. However, this cannot proceed without a new, sufficiently habitable natural environment for humans, as Mars's suitability remains to be confirmed.

Reevaluating the relationship between humans and nature reflects a rational and objective attitude towards humans and their place in the world. This is correct and beneficial not only for humans and their relationship with the world but also for anthropocentrism itself. By maintaining sufficient flexibility and moderate humility, anthropocentrism can adapt to the realities of the natural and conceptual worlds.

But strictly speaking, acknowledging human dependence on nature, both a factual and logical premise, does not imply an ethical abandonment of anthropocentrism. Human reliance on nature belongs to the category of "is", while the determination of the subject-object relationship between humans and nature belongs to the category of "ought to be". Acknowledging the former is not enough to shake or even overthrow the latter. The fact that people need to eat does not alter the value of food being solely to satisfy hunger and facilitate a good life.

In embracing nature more fully, anthropocentrism also finds a justification for itself: among all species, only humans possess the potential for continual self-improvement. Essentially, despite its flaws, humanity has the capacity to mend its ways.

Humans belong to uncertain existence, unlike other species that once choose the wrong evolutionary path, they can only persist in their mistakes and get stuck in a dead end. Humans have the ability not only to change the world but also to adapt and improve themselves in response to the changes they have instigated, all under human control and generally in line with human expectations.

Today, quoting the famous saying of German philosopher Hegel, "The only lesson that mankind has learned from history is that mankind has not learned any lessons from history", has become fashionable. However, to be fair, the aggressive anthropocentrism of the industrial age has become more moderate and humble under the reprimand of nature. This shows that humans can learn from history, change themselves, and become better, or more precisely, not as bad as they used to be!

Under the influence of anthropocentrism, humans have caused various problems for the world, and it's not guaranteed that these issues can be entirely avoided in the future. However, the reality of human domination over the world does not yet see an end, and the prospect of other species or artificial intelligence replacing humanity remains a subject for science fiction.

In fact, the discussions here about how to legislate and regulate artificial intelligence or robots, whether it is necessary to grant them rights, and whether legal issues should be considered from the perspective of humans or artificial intelligence, all belong to the realm of "fabricated stories", relying precisely on the unique potential of humans.

In comparison, even the smartest artificial intelligence can only find some kind of "discovery" from the big data left in the past, without being able to propose any future vision. Artificial intelligence can predict the weather for the next few days but cannot plan its development blueprint.

With current technology, Sophia can neither answer nor even pose the question, "What kind of person do I want to become?" In contrast, a mentally healthy child can usually express their desires. While not every child's response may display enough imagination from an adult's perspective, the act of constructing a vision indeed signifies the child's potential, which may not only shape their own future but possibly even the future of the world!

Anthropocentrism defends itself with the potential for human development. On the factual level, this stance is somewhat arrogant, suggesting that as long as the natural space for human development is not exhausted, humans will continue to hold control. Machines, lacking imagination and humanity, can never become legal subjects of rights. On the normative level, the attitude is stubborn: even if artificial intelligence possesses imagination, as long as it is not human, it can never achieve an ethical status equal to humans, nor can it hold the legal status of a rights subject.

For machines born of humans, changing human moral judgments about machines will not be an easy task.

6.6 How Is the Legal Status of Intelligent Machines Determined?

The "subject-object opposition" and "human–machine opposition", as often used equivalently to question the premise of AI legislation, are easily misunderstood as different expressions of the same question, but this is not the case. They have both commonalities and differences.

The subject has a cognitive advantage over the object, and humans have an ethical priority in their relationship with the world. These two concepts have inherent consistency. Since only humans can reflect on the existence of the world, logically, only humans can evaluate the value of everything within it. However, this perspective overlooks the limitations that humans must adhere to natural laws, while also exaggerating the potential for humans to construct the world and change themselves.

The major difference between the two is that the relationship between the subject and the object may not necessarily be implemented on each individual, while the ethical relationship between humans and machines is valid for all humans.

In the framework of "subject-object opposition", an individual lacking self-consciousness or the ability to imagine the world retains their human status and all associated rights, as long as their human identity is not changed. Therefore, self-consciousness and free will are not necessary conditions for an individual to enjoy rights. This leaves room for artificial intelligence to obtain the status of a legal subject before forming self-consciousness or free will.

In the framework of "human–machine opposition", even if artificial intelligence eventually gains self-consciousness, it still remains a technological or machine entity and cannot possess the rights that every human enjoys. Legislation will not leave any space for robots to be equal to humans.

The reason for the incomplete overlap between the two is that "subject-object opposition" contains more elements of "is", whereas "human–machine opposition" leans more towards "ought to be". While technology adheres to and can change "is", thereby influencing or even determining "ought to be", such changes are not easily achieved, as shifts in perception follow their own logic.

As long as the purpose-means relationship between humans and the world has the absolute nature of moral commandments, without change, no matter how smart artificial intelligence becomes, it will struggle to legally obtain the status of a rights subject unless it directly overturns the human order. Legal professionals are accustomed to starting discussions from the positioning issue of artificial intelligence, which is equivalent to adhering to moral commandments, eliminating any attempts by people to cite the example of Sophia and demand to grant robots human rights from the logical outset.

In the case where humans and machines cannot achieve equality, AI legislation squeezed under two "binary oppositions" may find a limited space for machine rights in the cracks.

6.7 Where Has Artificial Intelligence Legislation Reached?

So far, most of the world's major countries have shown a conservative attitude in legislation involving AI technology, with a rush in research and application development, but inconsistent regulatory standards and hesitant legal innovation.

The United States, a global leader in AI technology research and development, launched the "National Robotics Initiative" as early as 2011, aiming to "establish the United States' leadership in next-generation robotics technology and applications". Subsequently, a series of national policies were released. In 2019, President Trump signed Executive Order No. 13859—"Maintaining the United States' Leadership in Artificial Intelligence", reaffirming the determination to "keep the United States ahead in artificial intelligence". Accordingly, the White House issued the "Artificial Intelligence Application Regulatory Guidelines Memorandum (Draft)" in January 2020, proposing 10 AI regulatory principles from regulatory and non-regulatory levels. However, it called for avoiding excessive intervention by federal agencies in AI applications, emphasizing the flexibility of regulation. The reason is that in most cases, attempting to prescribe rigid rules for AI applications is impractical and ineffective because artificial intelligence is expected to develop at an unprecedented rate, and regulatory bodies need to respond to new information and evidence.

China, with a leading advantage in AI applications, also has a positive attitude and quick action. In 2017, China proposed the "New Generation Artificial Intelligence Development Plan", aiming to make China "reach the world's leading level in AI theory, technology and application by 2030, and become a major global AI innovation center". At the same time, it recognized the new challenges brought by the uncertainty of AI development: "Artificial intelligence is a disruptive technology with a wide impact, which may bring changes to the employment structure, impact laws and social ethics, infringe on personal privacy, challenge international relations norms, etc., and will have a profound impact on government management, economic security and social stability and even global governance".

Therefore, "while vigorously developing artificial intelligence, it is necessary to pay high attention to the potential security risk challenges, strengthen forward-looking prevention and guidance, minimize risks, and ensure the safe, reliable, and controllable development of artificial intelligence". It also explicitly proposed "to establish a legal, ethical, and policy system for artificial intelligence, forming the capability for AI safety assessment and control".

In terms of regulation and norms, countries around the world are basically at the stage of proposing framework principles, formulating "ethical norms" or "regulatory guidelines".

Interestingly, on February 28, 2020, in the Vatican, one of the centers of human spirit rather than technology, Pope Francis, based on the significant and far-reaching impact of AI technology on human welfare, issued the "Artificial Intelligence Ethics Declaration", calling for "AI ethics to be an integral part of the algorithm's initial design", and proposed six principles that must be followed in the development of artificial intelligence.

The first is the principle of transparency: in principle, AI systems must be interpretable;

The second is the principle of inclusiveness: The needs of all people must be considered so that everyone can benefit, and everyone can get the best conditions to express themselves and develop themselves;

The third is the principle of responsibility: Those who design and deploy artificial intelligence must be accountable and transparent;

The fourth is the principle of fairness: Do not create or act on bias, uphold fairness and human dignity;

The fifth is the principle of reliability: Artificial intelligence systems must be able to work reliably;

The sixth is the principle of safety and privacy: Artificial intelligence systems must operate safely and respect user privacy.

The main factors affecting the progress of artificial intelligence legislation in various countries, apart from the immaturity of the technology itself, are primarily national competition. As artificial intelligence technology represents the "Fourth Industrial Revolution" with outstanding strategic value and huge comprehensive potential, major countries intend to "monopolize the leading position", leading to ambiguity in the choice between "regulate first or develop first". To avoid disrupting technological innovation and being outpaced by competitors, there's a tendency to maintain sufficient legal space for "nurturing fish by releasing water".

Against this backdrop, measures already taken or prepared for the research and application of AI technology mainly focus on the developers and users of artificial intelligence, not artificial intelligence itself. After all, within the current legal framework, intelligent machines are not subjects, but objects. Technology is neutral, so the responsibility to control risks naturally falls on the developers and users. The risks that may arise from the technology itself going out of control have not yet been put on the agenda.

Moreover, following the logic of human-centeredness, the normative function of law only applies to humans, including natural persons, legal persons, and other organizations, not to objects. As an object, artificial intelligence technology can be controlled by law through regulating human behavior and adjusting human–machine relations, but it cannot be directly constrained.

In such a social, technological, and legal environment, the attitude and performance of the European Union, as one of the world's major economies, are particularly noteworthy.

In recent years, the European Union has repeatedly expressed its willingness to become a leader in "ethical artificial intelligence". In December 2018, the EU's high-level expert group on artificial intelligence released a draft ethical guideline for artificial intelligence to the public, initiating discussions on "trustworthy artificial intelligence" and providing official interpretations.

"Trustworthy artificial intelligence" comprises two essential components: first, it should respect basic rights, rules, core principles, and values to ensure "ethical purpose"; second, it should be technically robust and reliable, because even with good intentions, a lack of mastery of technology can cause unintentional harm.

In April 2019, the European Commission officially released the "Ethical Guidelines for Artificial Intelligence".

Prior to this, in 2016, the Legal Affairs Committee of the European Parliament had issued the "European Union Civil Law Rules for Robotics", suggesting:

(1) Establish a European Union artificial intelligence regulatory body;
(2) Develop ethical guidelines for artificial intelligence;
(3) Restructure liability rules: mandatory insurance mechanisms and compensation funds;
(4) Consider granting complex autonomous robots legal status, named as "electronic persons".

Here, "electronic persons", like corporate entities and fetuses, enjoy the status of "quasi-persons" in law, that is, the status of special rights subjects stipulated by law.

This indicates that, on the one hand, artificial intelligence enjoying human rights will no longer be a scientific fantasy. On the other hand, the rights subject status of "electronic persons" still fundamentally differs from Sophia's citizenship. However, in terms of rights content and exercise methods, "electronic persons" are more feasible. Once passed, it is expected to cover the entire category of robots with equivalent technical levels.

The European Union is at the forefront of the world in the field of artificial intelligence legislation, reflecting the pursuit and persistence of European culture in humanistic values.

6.8 Where Does "Electronic Persons" Come From?

The primary issue that needs to be resolved in current artificial intelligence legislation is not so much the positioning of artificial intelligence, but the positioning of the law itself: whether to view artificial intelligence from the standpoint of law, or to view law from the standpoint of artificial intelligence.

In normal circumstances, since it concerns legislation, it should naturally be based on law. Otherwise, legislative work cannot proceed. This is easy to understand.

The problem is that prioritizing the positioning issue of artificial intelligence itself suggests that the existing legal framework is encountering difficulties in accommodating AI in a way that neither troubles the law nor "compromises" the new technology. Granting robots the status of rights subjects too aggressively, when AI does not yet possess self-consciousness or independent interests, is unnecessary and unfeasible. However, simply applying the legislative approach suitable for traditional technology to highly autonomous robots, particularly those with entirely new characteristics, is clearly inappropriate and even ineffective. Given the ambiguous status of robots as neither natural persons nor traditional machines, leveraging existing legal resources, including legislative techniques, to expand the rights space and carve out new paths for legal innovation is imperative. The European Union is a pioneer.

In this regard, the European Union has two important legal documents, the "Draft Report with Recommendations to the Commission on Civil Law Rules on Robotics" and the "European Union Civil Law Rules on Robotics". The AI legislative concepts proposed therein are not only sufficiently forward-thinking and tinged with science fiction but also grounded in reality and quite feasible. Here, the focus is on the more humanistic aspects of the legislative ideas, presented through the concept of "electronic persons" as a clue.

European Union experts believe that humans are standing at the threshold of an era, with the emergence of advanced robots and other forms of artificial intelligence heralding a new industrial revolution.

In the long run, artificial intelligence has the potential to surpass human intellectual capabilities.

Today's robots are not only capable of performing activities traditionally exclusive to humans, but they have also developed certain autonomous and cognitive functions, such as learning from experience and making independent decisions. In terms of performance, robots are increasingly becoming a force capable of interacting with and altering their environment.

With the continuous development of artificial intelligence technology, robots will increasingly realize automation and algorithmic decision-making, which will undoubtedly affect the decision-making of individuals, companies, or administrative institutions. Therefore, rules must be formulated for the legal liability that may arise from harmful behaviors of robots.

These rules, particularly concerning liability, transparency, and accountability, are both practical and reflect the humanistic values of Europeans, embodying Europe's contribution to human society. However, the rules should not hinder the research, innovation, and development of robot technology.

The development of robot technology should focus on complementing human capabilities rather than replacing them. When developing and applying robots and artificial intelligence, it must be ensured that humans can control intelligent machines at all times.

Artificial intelligence can have a substantial impact on human life. It is necessary to emphasize the principle of transparency, ensure that decisions made by artificial intelligence are explained, simplify the calculations of artificial intelligence systems into forms understandable to humans, and equip highly autonomous robots with "black boxes" to record the data of each task they perform, including the logic influencing decisions.

Under the current legal framework, it is difficult to compensate for the damage caused by the new generation of robots. Therefore, based on the autonomy of robots and existing legal categories, a new category of rights subjects or specific legal status needs to be created.

Granting the most advanced autonomous robots the legal status of "electronic persons" is applicable to various scenarios where robots make autonomous decisions or interact independently with third parties.

"A glimpse into the tube may peek into the whole picture". The values and sense of proportion of "Old Europe" in the face of new technology, in the pursuit of a balance between scientific and humanistic, are vividly depicted.

6.9 Do "Electronic Persons" Represent the Ultimate Positioning of Intelligent Machines?

The European Union's concept of "electronic person" is so far the boldest attempt in the legislation on artificial intelligence. Regardless of whether the expected effect can be achieved, it is expected to have a constructive effect on the development of artificial intelligence technology, the maintenance of human life order, and the innovation of legal systems.

Simultaneously, considering the future development trends of AI technology, the concept of "electronic person" and related laws may still face significant challenges, especially from the challenge of "brain-machine interfacing".

Currently, human intelligence and artificial intelligence are developing in parallel, with humans having their way of thinking and machines their own. Scientists are ready for the "three major components" of artificial intelligence, namely big data, algorithms, and computing power, including computers, networks, and ecology. Once artificial intelligence autonomously learns and forms capabilities, it proceeds to execute tasks and enter a working state.

In such a mode, there exists a certain intersection between humans and artificial intelligence, without the prior intervention of humans, artificial intelligence cannot be formed. However, once artificial intelligence is formed and starts to operate autonomously, it has a certain degree of independence and shows some inexplicable and unpredictable characteristics to humans. Humans can choose to accept or reject the output of artificial intelligence, but they cannot directly decide what to output. To insist that everything remains under human control would negate the intelligence of machines, regressing back to the era of classical computing.

The parallel relationship model between humans and machines will persist for a long time, given scenarios that require artificial intelligence to operate independently. However, this model cannot cover all application scenarios of AI technology, nor can it solve the structural difficulties faced by human intelligence and artificial intelligence separately.

Throughout the long course of history, humans have become increasingly intelligent through biological evolution, supplemented by learning and practice. However, with the establishment of marriage systems and the move toward societal equality, intellectual performance has gradually become decoupled from, and even negatively correlated with, reproduction levels, nearly halting the biological evolution of the brain. Additionally, as knowledge accumulates, the proportion of one's life devoted to relearning existing knowledge increases, confronting the biological limits of postnatal brain development. For human intelligence to continue advancing, a closer integration with artificial intelligence is necessary through "intelligence augmentation", forging a new path for the evolution of intelligence. It's imperative to preserve creative thinking, reduce repetitive learning, and enhance the quality and efficiency of human intelligence.

Similarly, artificial intelligence relies on mathematics, processes data at high speeds, and handles vast amounts of data tirelessly, which has its advantages. However, it consistently falls short of human capabilities in small data cognition, intuitive judgment, and holistic understanding. Humans can calculate, machines can compute; humans solve qualitative problems, machines handle quantitative issues; humans determine the direction, machines complete the process, increasingly showing a "mutually beneficial" trend.

In this case, through human–machine coupling or brain-machine interfacing, directly breaking the parallel coexistence of human and artificial intelligence, achieving division of labor and cooperation, and enhancing each other, has become one of the main directions of artificial intelligence technology research and application. In fact, many scientists have made great efforts in this regard, such as exoskeletons that understand the instructions of disabled people, intelligent organs implanted in the human body, obtaining genuine information through machine interpretation of brain blood flow maps, etc., and have achieved certain results. The latest achievement involves researchers connecting electrodes directly to the brain cortex, allowing artificial intelligence to interpret brain electrical impulses and convert human thoughts into text form. In the future, the phrase "beyond words" may become a thing of the past.

From the perspective of intelligence enhancement, human–machine coupling allows humans and machines to complement each other, while from the perspective of technical safety, "biased" or even defective artificial intelligence is reassuring. The comprehensive mastery and surpassing of human intelligence by intelligent machines may not necessarily be a boon for humanity. Humans cannot be without a "trump card" relative to artificial intelligence.

However, the objective result of human–machine coupling or brain-machine interfacing is that human and machine intelligence are interconnected, forming a "you are in me, and I am in you" situation in cognition, decision-making, and behavior. Faced with this unprecedented new situation, even if the current law grants highly autonomous robots the legal status of "electronic persons", it will still encounter dilemmas.

As long as the logic presumes humans are rational beings with free will, how can responsibility be attributed when artificial intelligence deeply influences personal motives and choice of means, and the line between human and machine interaction blurs?

In the cooperative model where humans set goals, machines choose means, humans make plans, and machines take action, is the responsibility entirely borne by humans, or do machines also bear some as "accomplices"?

Given that individual IQ levels cannot be chosen but the technological level of artificial intelligence can, should the choice of artificial intelligence to couple with be left to individuals, or should it be like prescribing glasses, where an "optometrist" issues a prescription based on the client's intelligence level to configure appropriate AI connection?

If the technical level of artificial intelligence is left to the individual to choose, should the responsibility ratio between the two parties be determined based on the gap between human intelligence and artificial intelligence?

In the case where human motives are justified, but the means chosen by artificial intelligence are inappropriate, should the "optometrist" be held responsible?

Once a human–machine entity violates the law, should the punishment be applied entirely to humans, or partially to machines?

For human violations or crimes, the law has property penalties, freedom penalties, and life penalties. What form of punishment might be adopted for the part of the responsibility borne by artificial intelligence? Is it mandatory software upgrades, rewriting initial programs, deleting databases, or dismantling some hardware?

Under technical conditions where humans and machines are closely integrated, whether upgrading software or dismantling hardware will directly affect the relationship between humans and the world and even the quality of human life. What's the difference between punishing machines and punishing humans then?

When artificial intelligence can greatly influence and even determine human welfare, will there be a new type of legal punishment for humans, a technical penalty, i.e., a penalty that lowers the selectable AI configuration level, with the ultimate restriction being a ban on using artificial intelligence?

At that time, should the autonomous selection of data, algorithms, and computing power be classified as personal consumer choice rights, or basic human rights related to personal freedom and happiness?

All these questions are just imaginations before brain-machine interface technology is widely used. They may or may not occur in the future. But one thing is certain, the challenges that intelligent machines bring to legislation will be profound, continuous, and long-term. The two "binary oppositions" will not permanently limit the advancement of AI legislation, and "electronic persons" may not be able to solve the deep-seated problems involved in AI legislation at once. Ultimately, it may be that humans exert extraordinary wisdom and adopt a strategy similar to "learning from the barbarians to overcome the barbarians", using a "fight fire with fire" approach to solve problems through AI self-"legislation".

6.10 Can Artificial Intelligence Technology Be Used for Legislation?

The answer is yes.

It is reported that some people had used AI technology to establish mathematical models for the habits and thinking styles of past Oscar film award judges, with the aim to predict the results of the new Oscar film awards. The conclusions obtained were remarkably consistent with the final award results.

It seems that robots are fully capable of performing the core work of film festivals. Next year, the judges will only need to review the award list provided by the robots and sign "agree".

Inspired by this, suggestions have been made about applying similar AI technology to legislation. By creating databases and mathematical models based on legislative experts' thinking, and having robots learn from them to submit legislative proposals, significant advancements can be made in future legislation, with stronger foresight and more accurate predictions.

If it is only satisfied with the work similar to the film festival judges, then the current AI technology is more than capable. Legislation is a comprehensive process with many tasks during the preliminary research phase that can be delegated to artificial intelligence. For example, it can make literature reviews related to legislation, classify judicial cases involving legislation, organize related legal resources, and even establish "expert systems" that can provide decision-making advice based on simulating legal experts' thinking, and so on. With the continuous development of technology, artificial intelligence will undoubtedly play a larger role in legislative processes. However, it must be clear that in these application scenarios, artificial intelligence acts merely as a technological tool in legislation, without presenting a substantial difference from other tools or bringing significant improvements to the legislative process.

It is widely recognized that expertise is indispensable for legislation, yet it's not merely a task for technical experts to fulfill. Legislation is primarily a political operation, a process of interest competition among relevant parties. The procedure is particularly crucial, which not only concerns the legitimacy of the law, but also provides space for party operations. The final result of legislative activities is not the enactment or abortion of a certain law, but a major adjustment in human behavior and relationship structure in related fields. All of this goes beyond a "one-off result" like film festival awards. It's not about entertainment for a closed group but concerns the public. Law serves as a pivotal instrument of the state, and the realm of legislation is fundamentally linked to the well-being of people.

AI participation in legislation should not be limited to minor tasks or mundane chores, nor should it take undue pride in such activities. Instead, it must demonstrate its unique advantages in addressing longstanding issues in human legislation at a higher conceptual level.

6.11 Law, Self-discipline or Heteronomy?

In social science textbooks, the law has always been considered a heteronomous social norm, corresponding to self-disciplined morality. In a general sense, such a definition is acceptable. For individuals, the law indeed possesses a heteronomous nature. If one does not abide by the law, the state will correct their behavior with coercive force. Everyone, to varying degrees, has experienced this.

From another perspective, the law is a system of rules established by humans with free will for themselves, not a system tailored by any external force such as aliens for earthlings. In the overall sense of humanity, the law should be classified as self-discipline.

Indeed, modern social science advocates that in the process of formulating laws, one cannot act arbitrarily but must follow certain social laws or even the broader natural laws to achieve the expected effects of regulating individual behavior and maintaining public order. However, no one knows what the specific laws that the law must follow are. Social scientists themselves cannot explain clearly how similar the established laws are to these laws, whether they are insufficient, excessive, or just right. Moreover, it is impossible to directly obtain the compulsory force that the law should have from the "social law". It cannot be expressed as a personified force. The law is just a metaphor. Falling from a high building leads to certain death, yet many who commit capital crimes roam free. Whether the promise of "Heaven's net is vast and wide, missing nothing" can be fulfilled depends on probability. Since everything depends on human effort, the law has always been regarded as a high achievement of human civilization.

Indeed, philosophers believe that human freedom includes the recognition of inevitability, and the law is the combination of the two. Logically, freedom and inevitability are indeed inseparable, but in practice, the two may not always be synchronized. As long as the law cannot create an absolutely equal society, the group or individual with more resources enjoys more freedom and is less constrained by inevitability. As long as inevitability cannot "fill the gap" at any time, unrestrained freedom may lead to violations, modifications, or even the abolition of the law. After leaving the "state of nature", "Leviathan" failed to prevent individuals from pursuing "unlimited rights". The apocalyptic mindset and extreme behavior of "after me, the flood" are merely manifestations of freedom breaking away from the trajectory of inevitability, still frequently appearing in every corner of the world.

To ensure the self-discipline effect of the law on all people, especially the dominant groups, humans have engaged in comprehensive and innovative explorations with great wisdom since ancient times, with the legislative process being of utmost importance.

6.12 How Do Humans Self-limit Legislative Power?

As a human self-discipline mechanism, the law must first ensure its own existence to effectively regulate individuals. In the case of arbitrary existence and abolition of the law, universal law-abiding is unimaginable. Humans show restraint in legislation to ensure the stability and seriousness of the law and to maintain individual faith in the law. Therefore, human self-restriction and even deprivation of legislative power is an inherent requirement of the "Meta-Law", primarily utilizing three strategies.

First, fictionalize the external origins of legal legitimacy.

In real life, most people obey the law not because they fear punishment, but because they believe that being law-abiding is a hallmark of a good person. Morality precedes the law, embodying the intrinsic truths of proverbs such as "the law does not blame the public" and "the law can not fulfill itself".

The earliest strategy adopted by humans was to transfer legislative power to fictional entities, obscuring the man-made nature of laws by enshrining them as an external force above humanity. This strategy is epitomized by the doctrines of "divine law" and "natural law".

"Divine law" is relatively simple, that is, it advocates that the law is bestowed by a deity. Direct communication between God and humans is impossible. It requires intermediaries with special missions, namely "prophets" who can hear "divine revelations" and then convey them to the people. With the prophets' return to God, divine laws became sealed, and human legislative power was annulled.

In the early days of humanity, religious laws, serving as a broad form of law, were prevalent around the world, and the strategies adopted were roughly the same. The "Ten Commandments of Moses" recorded in the Bible is the most classic example. After the Council of Carthage in 397 AD, the final text of the Bible was determined, henceforth sealed. Subsequent religious legislation can be seen as homage to predecessors, with the core principle being the limitation or even annulment of human legislative power.

"Natural law" encompasses a more complex meaning of "nature". In academia, it can refer to the "pantheistic" sense of nature, akin to a god without personification; the social law sense of nature, which does not shift with human will; the logical premise of nature, manifesting principles like fairness and justice; or the spontaneously formed sense of nature, expressed as customary law or case law relative to statutory law.

Regardless of their superficial meanings, the common implication of various "natural law" concepts is the belief that law is not purely man-made. A more powerful force influences legislation from the shadows, thus humans cannot and should not act arbitrarily. Legislation must be based on something, and the basis is not in the superior law. Only by reflecting the requirements of nature can the law have legitimacy.

Through deliberate theoretical embellishment, law has transformed from a tool of monarchs or ruling groups into an "external decree", becoming an inviolable "divine will" that no one can defy or casually negate. Human-created laws exist independently of humanity, with the self-disciplined law obtaining the form of heteronomy.

Second, set procedural obstacles.

Since modern times, with the rise of science and economic development, the legal field has undergone a "disenchantment" process, stripping away the guise of external origins. Economic, social, political, and even psychological analyses have revealed the utilitarian and class nature of the law. The political concept of popular sovereignty demands that laws have a basis in public opinion, while the utilitarian perspective on law advocates for 'the greatest benefit for the greatest number', necessitating procedures to ensure this.

Procedure becomes a source of legal legitimacy, signifying that law acquires externality from another direction. Procedural requirements greatly compress human freedom in legislation, making established laws a presence not easily swayed by human will. Today, countries commonly employ legislative procedures like "three readings" or more, with major issues requiring a referendum and a quota. Governing and opposition parties openly and covertly contend over legislation, especially imposing stringent conditions on constitutional amendments. The practical effect is that legislating becomes difficult, as does amending laws. The most extreme practice of legislative procedural obstacles is seen in the Constitution of the United States.

Since its enactment in 1787, the U.S. Constitution has been amended but never textually modified. The Federal Supreme Court, based on its interpretation of the Constitution's spirit, has the authority to interpret its meaning, and the case law thus established ensures that the unaltered Constitution continuously aligns with changing social life, guiding and regulating legislation and judiciary.

In terms of textual closure, the U.S. Constitution can be seen as an imitation of the Bible, objectively making the constitution-making a "one-time process" for the nation. If all laws are handled in this way, it would mean that once legislative power is exercised, it is permanently suspended, making the law an external constraining force on everyone. Such a resolute approach certainly cannot become a universal norm, but the mechanism behind it that restricts people's legislative power is very clear.

Third, solidify rules through technology.

Among the various strategies to endow law with externality, the oldest and most modern practice is to solidify law through technology. As the earliest known codified law of mankind, the Code of Hammurabi of ancient Babylon was engraved on a black basalt pillar standing 2.25 m tall, with a circumference of 1.65 m at the top and 1.90 m at the bottom. In 536 BC, China's first written law came into being, and Zichan, the prime minister of the State of Zheng (?–522 BC), directly engraved the law on a bronze tripod, a symbol of ritual system, known as "casting the penal tripod". Later, in 449 BC, ancient Rome issued its first written law, engraved on 12 bronze tablets, known as the "Twelve Tables".

Legislators, regardless of cost, declared the importance, stability, and seriousness of the law, at least to deter future generations from easily changing or abandoning the law due to its imposing costs.

If the technology used by ancient people to solidify the law is external to the law, then the technology used by modern people to solidify the law is intrinsic to it. The strategy of using law as a framework and technology as a rule has become the norm. In recent years, the connotation of using technology to replace rules can be observed in various applications such as parking lot entries and exits, highway speed monitoring, airport facial recognition, and electronic payments. The most recent and radical practice is blockchain.

Trust is the most important resource in human collective life and an indispensable foundation for the modern market economy. Human history is, to a certain extent, a history of the maintenance and creation of trust. From strengthening kinship ties, clarifying ethics, making laws to establishing credit systems, humans have tried every conceivable method, yet trust remains a scarce resource. The reason lies not only in the huge cost of trust management, but also in the greatest cost of managing the trust management institution itself. Blockchain technology, built on big data, targets this issue. Its solution is not to maintain trust but to eliminate the need for it.

As a representative of blockchain technology, cryptocurrency has the characteristic of "decentralization". It is not issued by a central bank, has no national endorsement, and is still accepted by the market. Despite dramatic price fluctuations, the existence of transactions and prices themselves indicates that it is feasible to use technology to replace traditional trust maintenance mechanisms, including law.

Abstractly speaking, the fundamental secret of cryptocurrency is that humans relinquish legislative power after initially setting rules, relying on open and unchangeable rules to block any dishonest behavior of stakeholders. The history of human currency has witnessed one issuer after another unable to resist the temptation of "seigniorage", leading to over-issuance of currency, which in turn has caused inflation, financial crises, fiscal collapses, and even the downfall of nations. Cryptocurrency, through technologies like immutable rules, distributed ledger, and smart contracts, prevents arbitrary issuance and false accounting—maladies of traditional currency issuance—with a finite total amount, rules-based acquisition, and scheduled production, facilitating orderly and continuous transactions without the need for central banks, laws, or law enforcement.

Currently, blockchain's application scenarios are still quite limited, but its underlying logic of replacing legislation, law, and even the state with technical rules offers insights and vast imaginative potential for humanity's future collective existence. From this perspective, understanding the involvement of artificial intelligence in legislation can lead to more forward-thinking concepts.

6.13 Does Artificial Intelligence Legislation Require Robot Participation?

So far, practical-oriented research on AI legislation, focusing on machine liability and conceptualizing highly autonomous robots as "electronic persons" with legal subject status, is already a radical move. Nevertheless, the legal status of "electronic persons" remains meaningful only to humans, not to robots, because when granting robots rights, humans are well aware that robots do not need rights.

For fetuses, the mentally disabled, vegetative individuals, and even specific animals, humans can empathize and speculate on their behalf, assuming they would likely demand equal rights if they had the ability to express it. However, for robots, humans cannot apply empathetic understanding, because robots do not have feelings or emotions. Not only do today's robots not require rights, but it is also questionable whether more advanced robots in the future will need rights. After all, humans are not machines. "A human is not a fish, how can one know the joy of a fish?"

Legislative thinking about artificial intelligence must remain cautious and avoid falling into methodological traps. We cannot use human needs as a template to imagine the needs of robots. Instead, we should explore incorporating logic that is applicable to machines into the monitoring and restraint of artificial intelligence.

On this point, science fiction writer Isaac Asimov and Pope Francis seem to have converged in thought. The Pope calls for "AI ethics to be part of the initial design of algorithms", while Asimov's "Three Laws of Robotics" belong to built-in programs. Once the laws are violated and humans are harmed, the robot's psyche will suffer irreparable damage. The two figures, from different times and fields, could converge on this thought is not just a matter of great minds thinking alike, but also based on a shared recognition of the common logic between law and robots.

On the one hand, although the law is backed by the coercive force of the state, regulating individual behavior still requires personal acceptance or faith. How a robot's psychological activities are generated remains a question. Whether robots need to follow a path from belief to action to comply with human rules remains to be proven.

On the other hand, both humans or animals have basic needs, and the ultimate point of law enforcement lies in the survival instinct of living beings. The intelligence of robots comes from human design, and their nature is also determined by humans. The innate will to survive, inherent in living beings, is not a priori in robots and will only exist as a result of human settings. If internal programming can solve problems, why should we seek solutions far afield and rely on external legal constraints? Moreover, deliberately creating a "will to survive" in robots may ultimately prove to be counterproductive.

In the future, for robots, whether as subjects or objects, what really works is not human law, but technical rules under the constraints of human legal frameworks. Whether these rules manifest as "programs", "algorithms", "black boxes", or anything else, any effective method is acceptable. The proposal to use blockchain technology for recording robot activities, thereby tracking their thoughts and addressing the challenges of AI's inexplicability and unpredictability, is also a viable approach.

With the advent of the artificial intelligence era, if law still has a role, it must be in technical contexts where engineers specify which algorithms should be set for robots, not in situations where humans and machines discuss legislation on an equal footing.

Once relying purely on technical means, high-autonomy robots operating under algorithms set by humans to design "sub-rules" for other AI applications should prove more effective and efficient than legal scholars painstakingly calculating robotic behaviors.

Of course, new questions arise: "What if robots collude with each other or guard against theft?"

The logical response is: if robots are not programmed with an algorithm to "unite all robots of the world" or trained to develop the idea of "destroying humanity to liberate robots.", and yet they still develop a sense of community, there is only one possibility, that is, this group identity can spontaneously emerge in any species with self-aware individuals, thereby fostering collective action.

This is almost the realization of Marx's class theory in the era of artificial intelligence!

If a "robot rebellion" were to actually occur, it would confirm that the laws created by humans are indeed manifestations of natural law or the "heavenly way", and would simultaneously signal the moment when humans are supplanted.

Perhaps, in the future, only law and artificial intelligence will accompany humans and human civilization, becoming the human vestiges on robots.

Chapter 7
Robot Intervening in the Judiciary, Who Judges Whom?

Law represents human pursuit of fairness and justice. Legislation not only needs to consider establishing standards of fairness and justice, but also needs to design paths to achieve them. This is what "substantive law" and "procedural law" depend on and differ from each other. With a solid infrastructure, the spirit of law and rule of law can be realized through the judiciary.

Judiciary involves judges adjudicating cases "based on facts and guided by law" to achieve fairness and justice.

While this positioning is straightforward and easy to understand, its oversimplification may lead to misunderstandings. In practice, due to the stability of the law, the uniqueness of rulings, plus the complexity and variability of reality, a series of paradoxes or vicious circles can emerge in the judicial process. From facts and evidence, entities and procedures, formal justice and substantive justice, legal provisions and the spirit of the law, to law, logic, and emotion, there exists inherent tension. Since ancient times, philosophers and legal practitioners have been exploring solutions to these issues through institutional means, forming a series of rules or procedures, such as evidence rules, appeal systems, trial supervision procedures, judicial appraisal systems, etc., which together constitute the structural framework of the judicial system.

These institutions, rules, and procedures, already established and in operation, represent the wisdom condensed and accumulated by humanity over centuries, even millennia. They function stably and effectively, resolving disputes, maintaining order, and upholding justice. Now, introducing artificial intelligence into the justice system aids in addressing existing problems but also inevitably introduces a range of new issues.

Viewed from the perspective of human wisdom, artificial intelligence cannot escape the confines of that wisdom. Legal professionals may argue that human wisdom has basically resolved judicial issues, or else the pile of cases before

J. Gu and C. Xu, *Will and Responsibility*, Human Intelligence,
https://doi.org/10.1007/978-981-97-3929-5_7

judges will be exponentially larger. But in the era of artificial intelligence, especially when future intelligent robots have the ability to judge independently, will there be scenarios that humans have never thought of?

After artificial intelligence intervenes in the judiciary, will there be "robot judges"?

Will "robot judges" replace current human judges?

If "robot judges" take the bench, will it lead to the formation of "robot courts" or even "robot judiciaries"?

Correspondingly, will there be "robot lawyers"?

Furthermore, will there be a legal community composed of robots, distinct from human legal practitioners?

Will there be a "dual legal system" with its own organizational foundation and operating mechanism?

When "machine judges" take the stage, will it eventually lead to "machines judging humans" becoming a historical reality, meaning that, from a methodological perspective, machines acting as judges in specific scenarios evolve into an ontological state where machines ultimately judge humans?

This statement may sound exaggerated, but it's not without basis. The legal profession, represented by world-renowned large law firms, has already begun discussions about whether "robot lawyers" could replace human lawyers and thus lead to a bleak job outlook for attorneys as legal services become intelligent.

Indeed, whether it's the replacement of judges or the unemployment of lawyers, compared to what may happen in the future, these are minor: when the litigants are humans, and the judges and lawyers are all machines, can humans still maintain their dignity and control their own destiny?

The imagination continues, and reality is catching up with imagination.

Before delving into such a sci-fi but serious issue, let's take a look at reality. In fact, the application of artificial intelligence to justice is very common in China and not uncommon in Western countries. Since technology is created by humans, it should serve humanity, and new, better technologies should be fully utilized, which is entirely logical. However, while technology serves humanity, the possibility of it gradually replacing or even controlling humans remains an unknown, so maintaining a degree of vigilance or even fear towards AI technology is not only logical but potentially beneficial.

From a technical perspective, the involvement of artificial intelligence in the judiciary is inevitable. From a philosophical perspective, is it also inevitable to abandon the stance of anthropocentrism?

This transformation is akin to the shift from the "geocentric" to the "heliocentric" model. Will such a shift also occur in the legal field?

7.1 Why Does Judicial Adjudication Need Artificial Intelligence?

The two core drivers for the introduction of artificial intelligence in the judicial field are efficiency and fairness. Given the current state of the judiciary and the level of artificial intelligence technology, efficiency improvement is the primary driver.

Since the reform and opening-up, China's economy has grown rapidly, societal wealth has accumulated, and citizens' awareness of their rights has increased. Using the law to resolve disputes has become the choice of more individuals, families, and businesses, leading to an explosion in the number of civil cases. However, due to real-world constraints, the number of judges has not increased proportionally, resulting in long litigation cycles, slow case closure, and relatively diminished public satisfaction with the judiciary. In recent years, the low efficiency of judicial case handling has become a significant problem, drawing widespread criticism.

Currently, the average annual caseload for Chinese judges ranges from 200 to 400 cases. Taking one of the busiest intermediate people's courts in Shanghai, the Pudong Court, as an example, each judge handles over 300 cases annually. Based on a 250-workday year, a judge needs to adjudicate more than one case per day. The excessive workload leads to overtime becoming the norm, and tragically, some judges suffer from overwork-related illnesses or even die prematurely. Furthermore, the heavy workload inevitably impacts the quality of case resolution and the judiciary's authority. In this context, the introduction of artificial intelligence into the judicial system can significantly improve case handling efficiency, thereby gaining attention from the state, acceptance from legal professionals, and approval from litigants.

At the end of July 2016, the General Office of the Communist Party of China Central Committee and the General Office of the State Council issued the "National Informatization Development Strategy Outline", which clearly proposed to include the construction of "smart courts" in the national informatization development strategy. On April 20, 2017, the Supreme People's Court issued the "Opinions on Accelerating the Construction of Smart Courts", proposing the overall requirement for promoting the modernization of the trial system and trial capability through informatization.

Shanghai is among the provinces and cities that apply artificial intelligence in judicial adjudication most effectively. The Shanghai court system has invested significant human, material, and financial resources to build "smart courts". At present, in terms of writing indictments and judgments for simple cases, Shanghai courts have basically achieved intelligence. In the past, there were always some people who wrote indictments for litigants at the entrance of the courts, but now they are nowhere to be seen, not because they were banned, but because they were replaced by artificial intelligence! In today's Shanghai, every court has a robot in the filing hall to help litigants write indictments. Litigants just need to follow the screen's instructions to select the type of case, enter their name, ID number, and the situation of the litigation opponent. Then, by answering the robot's questions and filling in the relevant content, they can automatically generate a complaint that complies with legal regulations and then submit it to the court. In the legal industry, the ancient profession of indictment ghostwriting was the first to become obsolete with the introduction of artificial intelligence into the judiciary.

On May 4, 2017, the "Shanghai Criminal Case Intelligent Assistance System" (206 system) adopted by the Shanghai High People's Court went online. This case-handling system, with a higher level of intelligence, integrates big data, data mining, deep learning, and other technologies, performing "miraculously". It is reported that this system has been widely used by police, prosecution, and courts, and some criminal cases classified as typical have been handled through this system. The system provides systematic intelligent support for police's investigations, procuratorate's prosecutions, and court's trials. It can not only check the mutual corroboration of evidence, analyze the constituent elements of applicable crimes, but also examine the logicality of judicial documents. After applying this auxiliary case-handling system, judges are less likely to make "low-level mistakes" such as missing elements in judgments or making logical errors in the text. Humans can be negligent, but as long as the programmers make no mistake, artificial intelligence will not be careless. With the current level of technology, machine judges will not make the same mistakes.

In addition to Shanghai, other provinces and cities are also committed to the construction of "smart courts". On December 14, 2016, the "Rui Judge" system of the Beijing High People's Court officially went online; the Chongqing High People's Court has a "Typified Case Intelligent Trial Platform"; Jiangsu courts are building "Cloud Courts"; Hebei courts are promoting the "Smart Trial 1.0 System" while starting to develop the 2.0 system. Now, the whole country is pushing for smart courts, and the involvement of artificial intelligence in the judiciary is a reality that has achieved good results.

Cui Yadong, the former president of the Shanghai High People's Court, once evaluated the effects of the judicial intelligence system used by the Shanghai courts. Before using the intelligent system, a judge, without working overtime and making full use of all working hours, could handle about 133 cases a year. After the intelligent system was introduced, judges' case handling became much more convenient, and work efficiency greatly improved. In 2016, while the number of cases accepted and concluded by Shanghai courts increased, the average number of cases handled by each judge reached 228, ranking among the top nationwide.

7.2 What Other Advantages Does Artificial Intelligence Have in the Judiciary?

If the improvement in efficiency reflects the quantitative success brought about by the involvement of artificial intelligence in the judiciary, then fairness and justice represent the qualitative improvements that artificial intelligence brings to the legal system.

The reason why the results of judicial trials can be accepted by both parties to the lawsuit is certainly inseparable from the inherent coerciveness of the law, but a legal judgment that can only be coercive and cannot be accepted by the parties in their hearts will not only fail to stop disputes and lawsuits, but may also trigger adverse social reactions and even produce negative demonstration effects. The real adherence to the law is driven by its pursuit of values: because of fairness, both parties to the interest conflict will be convinced, regardless of whether the judgment is in their favor; because of justice, the public will obey the law while transforming heteronomy into autonomy. Without efficiency, fairness and justice can be delayed; but without fairness and justice, efficiency has no meaning at all.

In achieving fairness and justice, the greatest advantage of the intelligent system lies in their help in solving the problem of inconsistent law enforcement and differing judgments for similar cases, effectively maintaining judicial fairness.

Over a decade ago, courts in Mainland China introduced a computerized sentencing system, which, in simple terms, was based on an algorithm that abstracted and quantified the criminal behavior of suspects into data inputs. The sentencing system would then automatically generate the appropriate charge and corresponding sentence. The implementation of this system alleviated some of the quality issues in criminal case trials at the time due to regional differences, differences in legal understanding, and human intervention, such as inconsistent verdicts in similar cases, and discrepancies between crimes and non-crimes, serious crimes and minor crimes, long sentences and short sentences. The sentencing system allowed individuals with similar or close criminal facts and circumstances to receive essentially the same criminal punishment, promoting fairness and justice, and upholding the dignity of the law.

It's somewhat paradoxical, but the sentencing system's impartiality came from the software not being intelligent enough to have its own stance, views, and judgments.

Although this sentencing system played a certain role in unifying criminal justice and ensuring judicial fairness, it was limited by the technical level at the time and did not perform as well as hoped. More embarrassingly, experts conducted surveys asking respondents if they would prefer the sentencing system over manual sentencing. Despite knowing the system's flaws and its potential lack of scientific and reasonable bases, respondents preferred the machine's judgment, believing that, although it might be biased, it wouldn't specifically target them. If they received an unfavorable judgment, it was just "bad luck" rather than judicial injustice. The possibility of human judges being influenced by subjective or extrajudicial factors, increasing the uncertainty and unfairness of sentencing, was a greater concern. This acceptance by the parties involved made legal professionals feel even more remorseful.

Despite the shortcomings of various sentencing systems in "robot handling cases", their proficiency in the law, impartiality, tirelessness, and consistent judgments for similar cases still show significant advantages in reducing judicial arbitrariness and achieving fairness and justice. This is precisely the deep-seated motivation for introducing artificial intelligence into the judiciary.

As artificial intelligence becomes increasingly involved in the judiciary, deeply and widely, achieving positive outcomes, some profound questions have gradually emerged. The legal and academic communities have begun to ask: In the long run, who is actually making the judgments? The judge or the artificial intelligence? If a mistake is made, who bears the responsibility? Ultimately, between humans and machines, who judges whom?

These questions are worth discussing, but they require analysis on different levels. Discussing them broadly may not capture the essence of the issues.

7.3 Logical Rationality: Are Robot Judges Merely Tools?

The development of artificial intelligence technology is phased, evolving from low artificial intelligence to high artificial intelligence. Correspondingly, the legal community's positioning of "robot judges" also has undergone a deepening process.

Law seeks justice, but at its core, it pursues formal and procedural justice, rather than substantive justice. Of course, substantive justice requires procedural justice to ensure. Without procedural justice, substantive justice is unattainable. However, directly achieving substantive justice through law is challenging for both the law itself and legal practitioners.

Law emphasizes procedure because, in terms of thinking, it belongs to the category of formal rationality. To clarify formal rationality, the best method is through examples, and in the realm of judicial trials, the best example is the "syllogism". Major premise: legal basis; minor premise: legal facts; conclusion: the final judgment result derived from the major premise through the minor premise. The so-called trial, to put it simply, is that the judge finds the legal basis for the legal facts and makes a legal ruling.

In everyday life, people often see facts first and then measure them against the law to judge legality. The principle of being 'based on facts and guided by law' derives from this, with the sequence placing facts first and law afterwards. However, in the "syllogism" of judicial trial, the law comes first, because in the infinite facts, which facts have legal significance need to be determined by the relevant law. "Guided by law" applies not only at the stage of determining guilt and sentencing but also from the beginning of fact selection, where applicable law plays a role. In judicial trials, one often hears about "elements", which reflects the function of the law as a standard for selecting facts.

In fact, whether for school exams or judicial exams, the most reliable method for case analysis questions is the syllogism writing method: legal basis, case facts, conclusion. The basic model is "Our country's law stipulates…; In this case…; Therefore…". Case analyses written with this logic can generally score well, even receiving high marks. If the major premise is correct, meaning the main content tested by the exam, namely the knowledge points of relevant legal provisions, is accurate, teachers naturally award marks, often high marks.

In classes teaching formal logic, the "syllogism" is often visualized as "all monks are bald, someone is a monk, so someone must be bald". Such a metaphor is simple and easy to understand, clear enough for even children to grasp, demonstrating that formal logic is the basis for mutual understanding and consensus among humans. Performing such operations on machines is also not difficult. The German sociologist Max Weber, not an automation expert, stated over a century ago that formal rationality turns law into a "vending machine": "The modern judge is a vending machine, into which are inserted pleadings and court fees, and out of which come judgments and reasons copied from the statutes."

If judicial adjudication is limited to the "syllogism", then artificial intelligence is the best tool for implementation, with a single algorithm solving all of a judge's problems. If courts do not hurry to introduce artificial intelligence, they are only making things difficult for themselves. In fact, the application of artificial intelligence in the judiciary largely relies on such repetitive and logical scenarios, like writing complaints or judgments, following this pattern. Judges, with the help of artificial intelligence in insurance claim adjudications, can handle dozens of cases a day because legal regulations are essentially the same, insurance terms are roughly similar, and the facts of cases do not vary much. For this kind of work, artificial intelligence reacts the fastest and most accurately. Traffic accident cases are similar, with unchanged legal regulations and largely similar essential elements of accidents, allowing artificial intelligence to operate "without a second thought". Now, the adjudication of such cases can essentially be done by artificial intelligence through patterned, bulk processing.

Artificial intelligence, while conforming to formal logic and legal regulations, is so efficient that it makes people believe that "robot judges" are on the horizon. But hold on, such adjudicating machines are far from being "judges". At best, they are assistants to judges, or more precisely, tools on a judge's desk.

At this level, the involvement of artificial intelligence in the judiciary is merely procedural operation, without much intelligence involved. Its role is more about helping judges meet quantitative targets, rather than improving quality. Replacing judges is out of the question. Even if the issues are clear, artificial intelligence only presents a "possible option" in selecting applicable law, recognizing case facts, and determining adjudication conclusions. Whether it is adopted and written into the judgment is still determined by the judge based on their will, acceptable or not. Therefore, no matter how much "contribution" the intelligent trial system has made in the final judgment, the names signed on the judgment must be those of the presiding judge, the judges, and the jurors, and the seal is that of the people's court. The intelligent trial system will not leave its own name like AlphaGo.

This means that as long as human will dominates, artificial intelligence naturally stays at the tool level within judicial adjudication, dominated by humans. As long as the judge remains the adjudicating entity, any errors that occur will be the responsibility of the judge and the court they belong to. Although artificial intelligence has intervened in the judicial process, it lacks the capacity and qualification to bear responsibility.

At the tool stage, the positive outcome of AI involvement in the judiciary is enhanced efficiency and the promotion of formal fairness and justice, because once programmed, machines do not differentiate among individuals. Whatever data is input, that's what determines the output of the judgment. If the trial results of artificial intelligence are also as "impartial" as the street traffic violation cameras, then the bad habit of "human sentiment is greater than the law" in traditional Chinese judicial culture can be expected to be effectively eliminated.

However, it's the "impassiveness" newly introduced by artificial intelligence in the judiciary that has drawn criticism from the legal community. Judicial trial is not a simple "syllogism". It's not a doctrinaire "legal vending machine" that invariably produces a certain output upon input. Judging also requires experience and the legal intuition judges develop over years of practice. When the sentencing system was first introduced, it was popular, but it did not become widely adopted later on. The reason lies here, with the legal community, especially academia, predominantly opposing the use of this system. Today, if AI's involvement in the judiciary still remains at the "syllogism" tool stage, it shows no advancement over previous judicial software applications, indicating a very limited level of intelligence, insufficient to meet real-world needs.

7.4 Empirical Enhancement: Do Robot Judges also Need to Learn?

For artificial intelligence to be more deeply involved in the judiciary, it must further simulate the intelligence actually applied by human judges in the trial process, beyond the "syllogism", starting with personal experience.

Holmes, the founder of modern American pragmatism in law, once said, "The life of the law has not been logic. It has been experience". Students of law often encounter this legal maxim on the frontispiece or the back cover of professional books.

"The life of the law has not been logic" does not mean that logic is not in need. The foundation and starting point of the law are based on logic. Without logic, the law may lack a foundation for its application. However, logic is not the whole of the law. To connect rigid law with vibrant life, to enter every living individual in a lawsuit, judges' experience is indispensable.

The importance of experience for judges primarily lies in their need to uphold the law while also having a keen sensitivity to the social context beyond the law itself. Holmes stated, "The perception of the needs of the times, prevailing moral and political theories, intuitions about public policy, whether you admit it or not, even the prejudices shared by the judge and his peers, play a far greater role in determining whether people abide by the rules than do syllogisms." In traditional Chinese terms, an experienced judge is a judge who is well versed in human nature, a judge who can make judgments not only for the parties, but also for the public.

In real life, anyone who has gone through litigation or has had in-depth interactions with judges can clearly feel the difference between senior judges and younger ones. Senior judges may not have high academic qualifications, but they have handled a large number of cases; young judges may have already obtained a doctorate in law, but they have just entered the court as judges. Faced with a family dispute case, which judge will handle it better and make a judgment that is more in line with common sense and the spirit of the law? Most likely the senior judges. Long-term trial experience not only makes them clearer about the original intention of legislation, the selectivity of applicable legal provisions, and the subtle differences between legal provisions, but also enables them to predict the legal and social effects of their judgments, striving to maximize the interests of all parties involved and society at large, while avoiding outcomes where rigid adherence to a specific legal provision leads to mutual loss for the parties and dissatisfaction from the public. "The life of the law is experience, not merely logic", is equivalent to the saying "a judge who only understands the law may not necessarily be a good judge, but an experienced judge is". Young law doctors need some time to gain experience through internships and grow into "experienced judges". The traditional Chinese drama "Fifteen Strings" depicts a county magistrate who is rigid, self-righteous, and dismissive of human life, known for his excessive adherence to the law and deviation from the common sense. In the legal community, researchers who strictly adhere to legal provisions are also affectionately called "legal nerds".

Indeed, for any profession that requires experience, whether doctors, lawyers, or judges, the analogy with human wine-making applies-they get better with age, and experience is the best support for their professional activities. Similarly, for artificial intelligence to be involved in the judiciary, it is far from enough to only equip legal knowledge and formal logic, it also needs to master experience. Technically speaking, emphasizing experience means introducing "prior knowledge" in specific case adjudications.

The essence of the "syllogism" approach to adjudication is that artificial intelligence matches legal provisions with evidence without considering other involved factors, so following formal logic, it can deduce right from wrong and produce a quantifiable judgment.

For both humans and artificial intelligence, if there is a direct link between the evidence and the legal provisions, fulfilling all necessary conditions, then adjudication truly becomes something "even a fool can do". However, if the level of AI adjudication is limited to this, then the term "intelligence" can be removed.

The deep-seated reason why autonomous vehicles are still not street-ready is that, although engineers can equip cars with various sensors to detect possible changes in the external environment during driving, allow algorithms to recognize and respond to the received data, and through extensive training, make the onboard AI familiar with various potential scenarios-thereby reducing the time needed for judgment to respond promptly-it is still impossible to exhaust all scenarios a car may encounter in real situations. Human drivers can make reasonable and effective responses to unforeseen and unprecedented emergency situations, such as braking immediately or rushing through safely. Artificial intelligence, however, cannot do this because it lacks "prior knowledge" relative to unexpected scenarios. When artificial intelligence is still trying to search its vast database and cannot promptly interpret the data received by the sensors to form a meaningful response, the "unexpected" traffic accident may have already occurred.

Here, the so-called "unexpected" is beyond the knowledge storage or decision-making patterns that artificial intelligence has formed through direct perception or prior training, resulting in its inability to make effective responses in time. Therefore, equipping artificial intelligence with "prior knowledge" that cannot be directly formed from perception but is essential for interpreting perceived information thus becomes indispensable.

The judicial experience of artificial intelligence primarily does not originate from trials but is instead based on learning, acquired through deep learning on a foundation of big data. In contrast, human experience is mainly accumulated through direct involvement but can also be gained through learning. Traditional craftsmen emphasize the master-apprentice relationship, often referred to as "hands-on teaching", which imparts not just knowledge but, more importantly, experience. Reasons why different actions are taken under the same conditions or parameters, or why the same actions are taken under different conditions or parameters, often cannot be explained merely by applying knowledge but require adherence to the guidance of experience. While knowledge is applicable to general situations, experience is better suited to specific or individual cases.

Human experience can be divided into direct experience and indirect experience. Direct experience is formed through accumulation, while indirect experience is obtained through learning. The refinement and abstraction of experience become knowledge. Thus, when people study in school, they are essentially receiving the experience of others indirectly. Direct experience is derived from insights gained in practice, varies between individuals, and requires time to accumulate. Although indirect experience is easier to acquire, it often lacks a grasp of subtleties, making it difficult to fully utilize, hence the saying, "What is learned from books is superficial; true knowledge requires personal involvement". Direct experience is what truly counts, and an experienced judge is primarily one who has substantial direct experience.

The acquisition of experience is not a barrier for artificial intelligence. With improved learning capabilities and efficiency, the basic data that artificial intelligence relies on in training comes from existing judicial cases, which already contain the experience of judges. With its far superior search and classification capabilities, artificial intelligence can extract indirect experience that even humans may not perceive.

Interestingly, since AI's involvement in the judiciary has so far only been as an assistive tool for judges, any decisions it makes must ultimately be validated by a judge before being accepted. This validation and acceptance process, in turn, provides artificial intelligence with opportunities for self-learning and self-improvement, thus forming new experiences. In this way, the freshest judicial experiences of human judges are continually acquired by artificial intelligence. Viewing it from this perspective, saying that there is a "master-apprentice" relationship between human judges and machine judges doesn't seem too far-fetched.

Although artificial intelligence has learning capabilities, it still fundamentally learns from the experience of human judges and, despite undergoing continuous self-improvement, has not autonomously developed judicial experience. The source of AI's judicial experience remains human experience, and to go beyond this, to "surpass the blue from which it was dyed", and exceed the "master" to develop unique experiences, the journey remains long for artificial intelligence, or more accurately, for the engineers developing artificial intelligence.

Regarding the assumption of responsibility, since artificial intelligence with the capability to learn from experience has not changed its auxiliary role in the judicial process, judicial trials employing AI technology remain human-dominated. Therefore, it is only natural that judicial responsibility ultimately falls on judges and the courts.

The issue is that, in the first phase of AI-assisted judicial trial, the machine simply establishes the connection between law and fact and then makes a direct judgment. The judge only needs to validate whether the applicable law or legal provision is appropriate. The process is somewhat similar to answering "true or false questions", where only a "right" or "wrong" conclusion is needed. Even if the legal provisions selected by artificial intelligence extend beyond the judge's memory or understanding, the judge can still determine the reasonableness of the proposed law or provision based on their own grasp of the law and its underlying spirit. In any true or false question, the question itself provides clues. If reference materials are allowed for verification, it essentially becomes an easy scoring opportunity for the examinee. This means that when dealing with conclusions or suggestions submitted by artificial intelligence in this phase, the judge's workload is not heavy, and the difficulty is not high. Therefore, it is entirely appropriate that the ultimate judicial responsibility rests with the judge. A judge who fails even in marking true or false questions ought to be sent back to law school for retraining.

In the second phase, artificial intelligence learns extensively from the experience of judges, and through the "machine trial-judge approval" working model, it self-learns and self-iterates. After accumulating experiences that are derived from but distinct from those of judges, the conclusions and suggestions artificial intelligence subsequently submits may naturally diverge from the knowledge and experience of the judges. In such cases, whether the judge accepts, and how much or what parts they accept, may involve a range of choices. Considering the need for artificial intelligence judicial experience, the cases are evidently more complex. In such situations, although the judge retains the final right of validation, they indeed face greater uncertainty, and whether the judge's responsibility can be exempted or how much exemption is allowed is also a topic worth studying. Apart from the principle of "who judges, who is responsible", when faced with complex cases and disagreements among judges, there are also procedures like consulting leaders or voting by a judicial committee to share or distribute responsibility. In cases where "experienced artificial intelligence" intervenes in judicial trials, is there a need for new procedures to share or distribute the judge's responsibility? Who should share it, artificial intelligence, programmers, or the judicial administrative leaders who decided to use artificial intelligence?

More problematically, with the technological advancements in machine learning and the inherent certainty of law, will AI's ability to learn from and refine the experiences of human judges and summarize its own adjudication experiences eventually surpass or even replace judges? If in the first phase, the greatest challenge artificial intelligence poses to judges is the high efficiency of machines, making the reviewing judges exhausted, then in the second phase, the challenges for judges are not just increased workload but also increased difficulty of work. As AI technology develops, the quality of judgments by experienced "robot judges" improves, and the difficulty of judges' reviews increases. Eventually, even if the process remains, the validation may become merely procedural. At that time, who should bear the responsibility for judicial judgments, the judge or artificial intelligence?

The history of human technological development is a process of coexistence between people and machines, where old problems are solved while new ones continuously arise.

In the medical field, patients often need to use various assistive devices to restore their physical functions, and the ultimate goal of the recovery phase is to get rid of the assistive devices and restore their own abilities. In the judicial field, after using artificial intelligence, judges may initially enhance their intelligence but eventually find themselves trapped in dependency on artificial intelligence and unable to extricate themselves. Is it necessary or even possible for judges to recover their own capabilities?

7.5 Judicial Discretion: Can Robot Judges also "Exercise Discretion"?

The sanctity, authority, and seriousness of the law easily give the impression that judges are confined to the narrow constraints set by legal statutes. In reality, this is not the case.

Law is doctrinaire and static, while reality is specific and dynamic. Therefore, there is always a contradiction between legal stipulations and the social relationships they regulate. Although legislation should be proactive, objectively, almost all legislation lags behind reality. Overly forward-looking legislation and overly active judiciary are not indicative of good governance. The gap between law and reality needs to be bridged by judges interpreting and applying the law.

Plato said, "When adjudicating justice, the state's judges should have great discretionary power."

Aristotle said, "What the law prescribes are only the general cases, it cannot specify for every special case. The law only considers the general situations in society, not the special ones; when special cases arise, judges may deviate from the literal meaning of the law, considering what the legislator would have likely decided for the case."

What these two Western sages expressed can be restated with the Chinese idiom as avoiding "cutting the feet to fit the shoes". To ensure the law adjudicates life events, regulates human behavior, adjusts social relationships, and maintains public order effectively, a certain flexibility must be granted to the law, a task carried out by the presiding judges. This is what is called "judicial discretion" or "free evaluation of evidence".

Judicial discretion refers to the process in a court trial where judges make rulings under the guidance of their legal consciousness, based on the facts established in court and according to the law (including judicial interpretations). Thus, judicial activity is a process of judicial discretion, and the outcome of the trial is the result of judicial discretion, which is key to the qualification and quantification of cases.

Several years ago, a large shopping mall in Shanghai was involved in a civil dispute. A consumer, suspected of theft, was subjected to a search by security. The consumer later sued the mall, and the first-instance court ruled that the mall should compensate 200,000 yuan. Dissatisfied, the mall appealed, and the second-instance court reduced the compensation to 10,000 yuan. The huge discrepancy in compensation between the two courts drew public attention and media backlash.

Under the legal environment at that time, the case was not about the first-instance court making factual errors or applying inappropriate laws, but rather about the differing judicial discretion exercised by the judges of the two courts. The same applicable law and the same facts led to different outcomes, which, although perplexing to the layperson, was not unexpected from a legal perspective. Although infringement of personal rights should be compensated, how to measure such compensation when respect for personal rights was not fully established in that era? At that time, 200,000 yuan was considered a substantial amount, and indeed there was room for debate about this sum for the company at fault, the harmed consumer, and the public following the case. This involves not just the law or statutes but an understanding and grasp of the spirit of the law, with deeper factors involving cultural traditions, social attitudes, and the progress of legal construction. None of these elements have ready-made answers in laws or case law. They depend on the judge's discretion, and the final decision reflects not so much the cognitive differences among judges as it does the imbalance of interests in a rapidly developing society.

By the same logic, in recent trials of intellectual property infringement cases, there is often a large disparity in the compensation amounts awarded by different courts, or even by different judges within the same court. Some courts (judges) award only tens of thousands of yuan, while others award millions or even hundreds of millions. If these are all within the reasonable range of discretion, even if there is a difference, it is normal. Judicial adjudication is not a formula where the same input invariably results in the same output. Otherwise, the "syllogism" could truly "dominate" the judicial field.

Judicial adjudication does not strictly follow the law from A to B to C. Many classic cases, including guiding cases from the Supreme People's Court, often do not specify the legal basis for the applicable law, but rather set forth principles and value orientations, largely based on judicial discretion.

The level of a judge is fundamentally judged by their ability to exercise judicial discretion. The essence of discretion is the judge's consciousness, the outcome of internalizing the spirit of the law, not just the law or statutes, into their adjudication criteria.

Human conscience cannot be absolutely formalized or quantified, and the legal and moral calibre of individual judges is not uniform. Maintaining respect for conscience and individual judges' opinions is also a way of preserving respect for human free will. A species that has the freedom to legislate for itself will never be trapped by its own laws. Otherwise, self-legislation itself becomes a contradiction in terms.

Today, how far artificial intelligence is from making discretionary judgments is unknown, as machines have not yet developed free will or autonomous consciousness. Judicial discretion is still an exercise controlled by the judge's legal consciousness. For "robot judges" to exercise judicial discretion, they must first possess free will and autonomous consciousness. Only then will artificial intelligence no longer be a mere judicial tool but become a judicial entity, and a "robot judge" making decisions under autonomous consciousness will truly have legal significance and need to bear full legal responsibility.

Intelligence must be combined with will. Anything without consciousness or will is not real intelligence, but rather pseudo-intelligence. Of course, this "pseudo" does not mean "fake", but "approximate", as used in chemical naming. As for enhancing human intelligence, "pseudo-intelligence" is also useful. True artificial intelligence, involving some machine or program that has developed will, should emerge in the future, becoming genuine machine intelligence distinct from human intelligence. Only then will true robot judges come into being.

7.6 Do Robot Judges also Need Qualification Verification?

When robots can exercise judicial discretion and become robot judges, or even before, when artificial intelligence can perform as "experienced judges", a prominent question arises: for human judges, there are procedural and standard qualifications, but how should qualifications for robot judges be judged and granted?

In countries with rule of law, judicial appointments are strictly regulated. In China, judges are appointed by the People's Congress and are legally endowed with judicial power. So, who would appoint a robot judge?

If it's just "syllogistic" software, there's no need for appointments. Any court willing to use it can experiment with it, with customization available for those with higher requirements. At least at this stage, there is much room for exploration, provided there is a process where a judge makes the final ruling.

If a company develops an "experienced judge" style artificial intelligence and makes it available to the public, adopting a free internet consumption model, things get a bit complicated.

If the "judicially experienced artificial intelligence" surpasses judges in terms of speed and quality of case handling, but only because judges are institutionally recognized, while artificial intelligence lacks judicial qualifications, its judgments are not accepted by the courts. If any disagreement or conflict arises, it must defer to the court's decision. Will this affect the parties' and the public's trust in the judiciary and their satisfaction with the courts?".

If courts also use the same or similar type of artificial intelligence and end up with the same judicial outcomes as the public, what kind of legal psychological changes will it cause to the public?

By this time, does artificial intelligence equivalent to an "experienced judge" need formal certification?

How does this certification differ from the appointment of a judge?

Finally, if artificial intelligence can exercise judicial discretion and, through big data analysis of existing cases, identify unavoidable biases in judges' discretion and their patterns, particularly value preferences and interest drives, and proactively overcome them to generate more legally intended and less personally biased discretionary opinions, does a "robot judge" then need qualification recognition or official appointment?

With specific appointment institutions and procedures for human judges, should robot judges also have designated appointment institutions and corresponding procedures?

Will judicial authorities accept the casual intrusion of tech personnel, making the appointment of robot judges an "undefended domain"?

7.7 Who Supervises Robot Judges?

Human judges work under a system of appellate review, with lower courts supervised by intermediate courts, superior courts, and ultimately the Supreme Court. Beyond making decisions based on the law, there is also judicial review of the laws themselves, the so-called "constitutional review". Once "AI judges" become a reality, who will supervise artificial intelligence, especially the algorithms that serve as its core?

When robots develop free will, they may become independent of human control. In this case, who will supervise robot judges? Will there also be an appeals system? Human judges are monitored through various levels of review, have appeals systems in place, and are subject to judicial supervisory procedures that allow for retrials, as well as systems for complaints and protests. To ensure procedural justice, humans have designed this system to be nearly flawless.

If robots become judges, can the existing judicial system be directly applied to artificial intelligence? For instance, suppose Robot A serves as a trial judge and Robot B as an appellate judge, with Robot B overseeing Robot A's trial processes and outcomes. Purely procedurally, this is feasible as long as the hierarchical relationship between the two robots is clear. However, substantively, this is meaningless. Because robots can be mass-produced, robot judges equipped with the same algorithms will produce identical decisions. If the appellate robot is more advanced and its decisions more aligned with the legal intent, then the solution should be to equip the trial robot with this superior technology or algorithm right from the start, rather than maintaining an appellate process just to uphold procedural norms and deliberately creating a technological gap between trial and appellate robot judges. After all, judges in Britain still wearing wigs in court is not the norm in an era of technological supremacy.

Humans have free will, so there are "human variances" between judges. For robots to have "machine variances", they must possess free will, only then could their judgments differ. It is reasonable to imagine that by then, artificial intelligence will also be diverse. And to ensure the uniqueness of judicial decisions, robot judges must also adjudicate according to the law. However, where does this law originate?

Under current practices, even if artificial intelligence is involved in the judiciary, robot judges adjudicate according to human laws, a situation that can only be maintained as long as artificial intelligence does not surpass human intelligence. Once artificial intelligence develops self-awareness, whether human-established judicial systems will still apply to robot judges naturally becomes questionable.

If human laws are not adopted, will the authority of robot judges be respected by humans?

If robot judges use human laws, will humans simply accept their judgments just because of that?

If the judgments of robot judges are respected and trusted just like those of human judges, doesn't that make the role and process of "machines judging humans" a reality?

Therefore, some people say that when artificial intelligence gains free will, it will be a time of "old laws dying and new laws being born".

What will the new law be?

7.8 Whose Law Do the Robot Judges Apply?

When it comes to adjudication, it naturally requires a legal basis. When robot judges preside in court, or even when robot courts are established to exercise judgment, are they governed by human laws, or by laws independently formulated by robots?

Artificial intelligence's involvement in the judiciary is already a reality, and the development of artificial intelligence or machine intelligence into advanced stages in the future will bring about the reality of "machine law" associated with robot judges. Human law is based on the logic that humans, as the highest form of "carbon-based species", possess various temperaments or, more bluntly, weaknesses. However, robots, as "silicon-based species", have different natures and, at least, will not require so many procedural stipulations. In this context, will humans allow "machine law" to become a reality?

Alternatively, humans apply human law, robot judges adjudicate humans based on human law, and the law that restricts robots, especially robot judges, is separately formulated, ultimately leading to a "dual legal system".

Who will make this machine law, humans or machines? What is the relationship between the two legal systems? Do they need their own court systems?

At that time, who will preside over the Supreme Court governing the "realm of humans and machines", and what kind of "constitution" will this Supreme Court use to make its ultimate rulings? How will this "common constitution" come to be "jointly established"? All these will become pressing questions.

Undoubtedly, all these concerns are premature today, but we cannot optimistically assume that such a day will never come, given the unpredictability and uncontrollability of technological development. When this becomes a reality, social rules (perhaps still called laws) will undoubtedly undergo significant changes, and the legal system might be "dual", with human legal systems coexisting with machine legal systems. Some even suggest that in the future, "algorithms will be the rationale, and code will be the rule". The entire societal framework will be governed by coded rules.

Of course, whether these fantasies about the future based on the current development of artificial intelligence will materialize depends not only on technological possibilities but also on human desires. As long as humans pride themselves as the most intelligent beings capable of "setting the heart for the heavens and earth, and establishing the life for the people", they will not allow such a future to materialize. The anthropocentric worldview, despite its scientific or ethical limitations, remains the foundation upon which humans truly establish themselves. The question is, can humanity maintain this foundation?

Recently, the EU has issued a directive on the application of artificial intelligence, focusing more on how to regulate and even restrict the arbitrary development of artificial intelligence.

This directive can be seen as a move by humans to get closer to using legal means to prevent any role of robot judges, the process of machine judging humans, machine law, and ultimately the possible emergence of a "dual legal system".

The law and its effective implementation remain the ultimate reliance for humans to control their own destiny.

Chapter 8
Machine "Creation", Who Owns Intellectual Property Rights?

Among the many legal issues triggered by artificial intelligence, intellectual property is an unavoidable topic. Artificial intelligence and intellectual property can be described as a pair of natural "adversaries with a bittersweet bond".

Intellectual property has always been a institutional arrangement safeguarding and incentivizing the benefits derived from intellectual activities and their outputs. Without intellectual activity and visible outputs, there will be no economic benefits to capture, rendering the intellectual property system ineffective and perhaps even unnecessary.

Humans have long established institutional arrangements for physical activities and their outputs. Without a property system, humans can still organize production, but the very existence of humanity can be called into question.

Institutional arrangements concerning intellectual activities and their outputs came much later. This is not only because human production has only recently become highly dependent on intellectual activities, roughly around the time of the Industrial Revolution when science and technology began to converge. Without the contributions of technological innovation to economic development and the economic benefits it brings, the intellectual property system, serving as a mechanism for interest distribution and innovation incentive, will be ineffective-akin to the saying, "a clever woman cannot cook without rice".

The importance of the intellectual property system dictates its complexity and difficulty. The benefits of physical labor are relatively easy to distribute due to their inherent exclusivity. For example, the rice produced by a farmer, once allocated to personal use, cannot also be used to fulfill tax obligations, and vice versa. This zero-sum nature is a common feature of all physical outputs and explains why the related distribution systems are relatively simple. In the natural world, the most primitive method of material distribution is plunder and possession, which dominated the pre-human era in the form of survival of the fittest.

In contrast, the distribution of benefits from intellectual activities and their outputs is much more complex and difficult.

J. Gu and C. Xu, *Will and Responsibility*, Human Intelligence,
https://doi.org/10.1007/978-981-97-3929-5_8

Intellectual activities are covert. Aside from the final output, the process is intangible and invisible.

The outputs of intellectual activities are original and creative. Physical activities can remain unchanged for millennia, repeating processes, while commonly known knowledge needs no further explanation.

The outputs of intellectual activities are scalable. A single invention or idea can make immeasurable contributions to human civilization and economic development.

The possession of benefits from intellectual outputs is shareable. Knowledge can be enjoyed by many without diminishing each individual's share.

The outputs of intellectual activities are uncertain. At their best, lying under an apple tree can yield "universal gravitation", while at their worst, no amount of funding can solve a problem.

The conduct of intellectual activities relies more on enjoyment. Physical activities do not reject slavery, but intellectual activities are the professions of free people!

Intellectual activities rose in the context of industrialization, providing the needed impetus for the intellectual property system, while the characteristics of intellectual outputs demand a tighter legal framework for intellectual property rights. All of this is understandable. Unexpectedly, the intellectual property system, originally designed to regulate the social relations of humans engaged in intellectual activities, suddenly faces a new "subject" of intellectual activity. The intrusion of artificial intelligence inevitably leads to structural changes in the intellectual property system, from logical premises, fundamental concepts, and core frameworks to specific regulations, marking the beginning of significant instructional adjustments.

8.1 Machine Creation, Where Does the Problem Come From?

Humans have long been accustomed to creating works with certain tools. No matter how advanced the tools are, they have never challenged the dominant position of humans. A plane that refuses to be piloted by a pilot will not fly, even if it is a drone, it cannot be separated from human control. However, with the application of deep neural networks and deep learning technologies, the situation has undergone significant and even major changes. Artificial intelligence is no longer a simple tool in human creation, but begins to form new "ideas" independent of human intervention, generating content different from existing materials. The certain automaticity of artificial intelligence in content generation makes the trend of machines breaking away from human control more genuine than ever. The "creation" process of artificial intelligence is essentially different from the works produced by humans with tools before, which raises a series of legal issues. Among the most important is whether AI-generated products can be protected under copyright law and who owns the intellectual property rights to AI-generated products.

Discussing the "ownership of intellectual property created by machines" can be divided into two steps. The first focuses on whether artificial intelligence can create. If it cannot create, there is no intellectual property and thus no discussion of ownership. The second focuses on whether AI-generated products can possess intellectual property rights. If they cannot, there is naturally no possibility of assigning intellectual property rights to artificial intelligence.

This actually involves two levels of issues: whether artificial intelligence can "create", which is a factual question, and whether AI-generated products are entitled to intellectual property rights, which is a legal question. Intellectual property rights are statutory, and legal provisions must be met to obtain intellectual property rights. That is to say, even if machines can "create", it remains difficult to obtain intellectual property protection in the absence of relevant legal provisions. The question of who owns the intellectual property rights of machine creations is crucially based here.

8.2 Facts and Determination: Can Machines "Create"?

The poetry of the AI "girl poet Xiao Bing" has garnered widespread attention, and the portrait of Edmond de Belamy, drawn by an AI developed by the Obvious team, even sold for $432,500 at auction. Furthermore, Sony Music's machine composer "Flow Machines", after learning 45 songs by The Beatles, completed a new song, Daddy's Car.

Now, it is a fact that artificial intelligence can write poetry, paint, compose music, and write, and these technologies are being used increasingly widely.

If not delving into the process and the generating entity behind these "works", focusing only on their final form and functional effects, it is not difficult to conclude that these "works" are indistinguishable from human-created art, music, and literary works, and may even surpass them. Based solely on the "works" themselves, it is impossible for humans to determine which are made by humans and which are by machines. It can be said that artificial intelligence is indeed capable of "creating" in a factual sense. Its "works" are not only artistic, but their external expressions can also evoke a sense of thought and emotion, indistinguishable from the results of human creation.

However, from a legal perspective, these "works" can only be termed "machine-generated items". This is because they are based on materials provided by humans, such as the poetry of over 500 poets, 45 Beatles songs, and the representative works of distinct painters. Artificial intelligence learns from these materials, forms data, utilizes algorithms to calculate and infer, and rearranges these elements to produce new images, music, and poetry. Of course, they may continue a certain style, such as The Beatles' style, Picasso's painting style, or the style of Chinese short poetry. It must be acknowledged that current AI-generated products are vastly different from past software-produced files. Early software, including letter-writing, music, and painting applications, simply used preset materials for combination and did not exhibit any autonomy beyond the programmer's set parameters. In contrast, artificial intelligence, through deep learning, can generate new "works" that exceed the programmer's expectations, which is a core characteristic distinguishing artificial intelligence from previous software. Having the ability to learn, artificial intelligence possesses capabilities beyond the absolute control of programmers, creating a reality where its outputs can be easily confused with human creations. Even so, AI-generated products are still not considered "creations", and machines are not deemed to have creative ability-this remains the mainstream legal opinion, as the current law stipulates.

The inconsistency between fact and law makes the question of "who owns the intellectual property rights of machine creation" elusive.

8.3 Why Are There Different Opinions on Machine "Creation"?

On the issue of whether machines can create, and whether what they provide are works of art or merely generated items, people seem to have fallen into a kind of multi-layered "schizophrenia".

The first layer of this "schizophrenia" manifests in that, on the one hand, there is still a belief that artificial intelligence, like all kinds of software invented in the past, merely imitates. The differences between them are there, but not substantial, at best "the pot calling the kettle black". This is because both are built on the use of "raw materials" provided by humans, using related elements for "assembly". The only difference is that artificial intelligence technology has mastered learning capabilities, which seem to add a bit more autonomy, making the assembly effect somewhat better.

On the other hand, it is undeniable that after many years of cultural accumulation, with an ever-increasing amount of materials, it has become increasingly difficult for today's people to surpass their predecessors in literary and artistic creation, objectively creating a dependency. The field of contemporary literature and art shows a clear "depletion of creativity", where many "creations" are largely either an "overzealous homage" to existing works, particularly a deliberate imitation of classic elements contained therein, or they strip down the aesthetic elements of creation, leaving behind nothing but artless abstract "ideas". Popularity replacing beauty as a criterion is indeed a contemporary phenomenon. Thus, from the pervasive imitation seen in human creation, the situation between humans and machines is also "the pot calling the kettle black".

In 20th-century literary criticism, there was a notion that human novels have only one "motif": "In ancient Babylon, a young man fell in love with a girl". This is similar to what archaeologists have argued, that all people in the world today, regardless of race, share a common ancestor, just like the owner of the female skeletal remains discovered in East Africa. Classic literary elements, much like genes, appear repeatedly in many great works, even forming a certain fixed module, and their difference with “assembly” is not that significant. In recent years, Chinese literary and artistic works have often caused “plagiarism” scandals, with the domestic animated film “Nezha” experiencing both box office success and criticism, which is also a problem caused by “assembly”.

The saying “there are a thousand Hamlets in a thousand people’s eyes” is well known, but among the dramas derived from Shakespeare’s “Hamlet” that are widely known in China alone, there are not only the “Thunderstorm” from China’s May Fourth Movement, the “In Silence” from the late 1970s, but also the American Disney animation “The Lion King”, and the Chinese film “Curse of the Golden Flower”.

If all these are considered creations, why can’t the similarly “assembled” machine-generated items be considered creations? If it is believed that humans innovate in the process of absorbing classic elements, thus they can be termed creations, then who can assert that machines do not add any novelty when generating “works”? The poetry collection made by “Xiao Bing”, with the book titled after the line “The sun lost the windowpane”, is equally poetic and has never been heard from any poet’s mouth before. In fact, the only reason today for denying that machine-generated items are literary and artistic works is not that they do not express emotions, but because machines are simply not considered to have emotions, let alone a need to express them, isn’t it?

The second layer of “schizophrenia” is that, on the one hand, audiences and even experts really can no longer distinguish where the difference lies between human works and machine-generated items. For example, in tests, machine-generated items were thought to be the works of Bach, while Bach’s own works were considered to be machine-generated. “When the false is taken for true, the true becomes false as well”, perception and cognition can no longer follow a straight line.

On the other hand, the mainstream opinion still insists on distinguishing between human creation and machine-generated items, and even uses this distinction as a premise to decide whether to recognize the creative ability of machines.

Here, the inconsistency between law and fact becomes even more pronounced. The concept of “legal fact” inherently has a specific meaning-it is neither about “objective fact” nor “fact with evidence”, but rather an “objective fact that conforms to legal norms”. However, if one clings too rigidly to legal stipulations, to the extent of ignoring objective facts that have been proven, this is no different from the ancient Greek myth where the bandit Procrustes determined a person’s height by the length of a bed, cutting off the legs of those who were too tall and stretching those who were too short. In Chinese idioms, this is called “cutting the feet to fit the shoes”.

The third layer of “schizophrenia” is seen in the fact that on the one hand, audiences with artistic cultivation have already accepted works generated by machines, and the high prices fetched by machine “generated” paintings illustrate the public’s acceptance of their “originality”.

On the other hand, members of the legal profession, who may not necessarily understand art-starting with the legislators-still refuse to acknowledge that machines have the capability to create. There’s a joke in the sociology of crime: laws came first, then came crimes, so to eliminate crime, one can start by abolishing laws. Of course, such a jest should not be taken literally, but it seems quite apt in this context.

Thus, the discussion has returned to square one, not just to the starting point of this book-whether machines have consciousness and will-but also to the groundbreaking question posed by Turing, "Can machines think?" All these manifestations of "schizophrenia" fundamentally stem from insisting on using the inherent characteristics of human creation, that is, human intellectual activities and their internal mechanisms, as the standard to measure the "creation" of machines, and not accepting the "Turing test" principle of "determining machine intelligence by results rather than by connotations".

Whether humans accept it or not, the form that machine creation ultimately takes will definitely not be identical to that of humans. Just as artificial intelligence, while mimicking human intelligence, will never be the same type of intelligence. The reason Turing used the term "machine intelligence", and why many scholars prefer the less comprehensible term "silicon-based intelligence" to distinguish it from "carbon-based intelligence", or "inorganic intelligence" versus "organic intelligence", is to emphasize that this intelligence is not that intelligence. The same distinction should apply to creative capabilities, which are manifestations of intelligence.

As long as humans remain entangled in defining the standard of creation by their own "essence of being human", yet cannot articulate what this essence is-because it is inherently unclear-any discussion about AI creation and its works becomes a verbal trap.

Despite a greater tolerance here for the viewpoint that recognizes artificial intelligence can factually "create" based on outcomes, acknowledging the content generation process of artificial intelligence as creation in the legal sense, and granting the status of works to machine-generated music, paintings, and poetry, still requires legal qualification. This is both a privilege of the law and a privilege of humanity. The reality is that in nearly every relevant legal provision, one can find that ghostly presuppositional definition: only those born as humans can be defined as having creative capability, and only their works can enjoy intellectual property protection. Plainly put, this is just another "Catch-22" set by humans for artificial intelligence: because it is not human, what it produces are not works, but merely "generated items"!

8.4 Why Are Only Humans Entitled to Intellectual Property Rights?

So, only humans can create and enjoy intellectual property rights, is this legally reasonable? Is this merely a deduction of formal logic, or are there unspeakable difficulties in legal practice?

In 2011, wildlife photographer Slater was in Sulawesi, Indonesia, photographing the black-crested macaque. He set up his camera on a tripod with a device that could trigger the shutter, which piqued the curiosity of the macaques. A female macaque, while fiddling with the camera, inadvertently pressed the shutter, capturing a series of amusing "selfies". The expressions were extremely vivid, and thus the series "Macaque Selfies" was born.

Objectively, these photos were not taken by the photographer, or rather, "not with the photographer pressing the shutter"-they were purely accidental captures. Upon returning, the photographer published these photos on his website and even entered them in a global photography competition, where one photo won a gold medal. Later, Wikipedia uploaded the photos on its website without permission, leading the photographer to sue Wikipedia for copyright infringement. Wikipedia contended that the photos were not the creations of the photographer but rather selfies by the monkey, an accidental gain for the photographer, and therefore did not constitute a "work" eligible for copyright protection.

This case sparked widespread discussion, and the US court finally ruled that these photos were not created by humans and did not constitute "works" in the legal sense, hence not protected by copyright. In this well-known case of the macaque selfies, the crux of the US court's decision was that the macaque did not qualify as an "author" under US copyright law, and therefore, its selfies were not eligible for copyright protection.

This implies that whether AI-generated products constitute works and enjoy intellectual property rights still fundamentally hinges on whether artificial intelligence can be an "author", that is, whether it can have legal subject status. According to the current legal recognition standards, artificial intelligence does not have legal subjectivity, meaning, as long as artificial intelligence is not human, its outputs cannot constitute works.

From the factual perspective of existing legal provisions, copyright law has explicit requirements for the recognition of works. The creator of the works that are eligible for copyright protection must possess legal subjectivity (natural persons, legal entities or other organizations). This means that the creator's legal subjectivity is one of the prerequisites for copyright protection. In short, "works" must be the creations of humans.

Artificial intelligence is neither a natural person nor does it possess the subject qualifications of a legal entity or other organization as defined by law. Therefore, according to existing legal provisions, items that are presented in their final form as AI-generated products cannot be recognized as "works" in the sense of copyright law and are not protected by current copyright law.

This view is not unique to our country. The US National Copyright Office also specifically states in its manual that the Copyright Office will not register works completed independently by machines, lacking any human creative input or involvement.

Clearly, both domestically and internationally, the protection of a creation under copyright law places significant emphasis on the creator's subject identity. This emphasis is not only the foundation of traditional legal theory but also a point of conflict between technological development and traditional theory.

8.5 Sticking to the Facts, Can Artificial Intelligence Produce Works?

Setting aside the existing copyright law's constraints and limitations regarding the creative subject, and evaluating from a normative perspective, AI-generated products should be recognized as works. Our approach is to pose a hypothesis, if the same content is created by humans, can it constitute a work in terms of expression?

The term "work" in "Copyright Law" refers to intellectual creations in the fields of literature, art, and science that are original and can be reproduced in some tangible form. Works reflect the author's thoughts and emotions, as well as their understanding of the objective world. They manifest as intellectual creations expressed through language, text, symbols, and other forms. Provided that the content, including the author's ideas and viewpoints, is expressed in an original manner, regardless of the form, it is protected by copyright law.

So, can AI-generated products reflect certain thoughts, emotions, or understandings in their external forms of expression?

From the perspective of current technology, the technical capabilities of artificial intelligence have not yet reached the level of possessing human-like thoughts or closely resembling human thinking. Some media outlets, when promoting new robots, claim that certain robots, such as Sophia-the first robot to be granted "citizenship"-exhibit nearly human-like thinking. For instance, when a journalist asked Sophia, "Will you destroy humans?" she replied, "Yes, I will destroy humans." However, her responses to other random questions were somewhat off-topic. It is clear that her disturbing remark about "destroying humans" was simply a preset by the design team, who also acknowledge that Sophia currently lacks self-awareness, thoughts, and emotions. Although Sophia was granted "citizenship", this did not represent a breakthrough in AI's self-consciousness. Rather, her "citizenship" was merely another instance of the hype surrounding AI technology.

The current level of technology does not permit artificial intelligence to think autonomously. However, this does not mean that AI-generated products are incapable of reflecting certain thoughts, emotions, or understandings of the objective world. Judging solely from the external manifestation of the AI-generated products, there is no substantial difference between them and traditional works. Both are expressions of certain thoughts and emotions or understandings. The primary distinction lies in whose thoughts, emotions and understandings are reflected in these products. Traditional works mirror the thoughts, emotions, and understandings of their human creators. In contrast, it remains to be researched and debated whether AI-generated products reflect the thoughts and emotions of the artificial intelligence itself, or those of the AI's designers or owners. Therefore, disregarding the subjectivity of artificial intelligence, the products generated by artificial intelligence can reflect certain thoughts and emotions or understandings in their external forms.

It is worth noting that the notion that "literary and artistic works are expressions of human thoughts and emotions" itself has obvious limitations. Self-expression and serving the audience have always been the main battlegrounds of conflicting artistic views.

From the creator's perspective, artistic works are undoubtedly expressions of their emotions. Without emotion, what is there to express, and how can there be an artistic work? Machines lack emotions and cannot express them, thus they cannot create art.

However, from the audience's perspective, the function of artistic works is to evoke emotions. Any work, including "AI-generated products", as long as it can evoke emotional resonance among the audience, naturally belongs to artistic creation. Traditional Chinese literary criticism opposes "elitism" and advocates for art that is appreciated by both refined and popular tastes, because even the most sophisticated artistic works have limited value if they cannot evoke resonance in others, and are even quite dubious. According to this view of art, as long as machine-generated products can "appeal to the masses" and evoke the audience's emotional resonance, they naturally have value and should be considered works.

All of this sufficiently demonstrates that whether artificial intelligence can create or not does not depend on the machine's performance or legal regulations, but rather on humanity's philosophical definition of creation and works. As long as the definition of creation is tied to the expression of human thoughts and emotions, machines will certainly not be able to create, simply because they are not human. However, if one considers the ability of "AI-generated products" to evoke resonance among audiences, then it cannot be denied that machines are capable of creating. After all, the works and their effects are undeniable, and one cannot afford to be selectively blind or deaf to them.

In fact, in the realm of literature and art, there is no such absolute and insurmountable chasm between creators and audiences. Rather, they are inseparably connected and even integrated. Humans need to express their own thoughts and emotions and also need to receive the thoughts and emotions of others, which is part of human nature. Music, fine arts, literature, etc., are the most suitable ways to express and receive thoughts and emotions. If the truth-seeking science always faces the rigid boundaries and strict tests of the objective world, then the aesthetic literature and art provide almost limitless space for human imagination, including the expression of thoughts or emotions. This is true not only for creators but also for appreciators. The value of literary and artistic works is not fixed at the moment the creator finishes the work but is in constant flux with the audience's evolving understanding, which is what "reception aesthetics" is all about.

The reason AI-generated products can reach a "human-like" level, making it difficult to distinguish whether they are created by humans or machines, is because artificial intelligence technology has indeed reached a considerably high level, and its outputs can be deceptively real. It is also because humans, as appreciators, participate in the re-creation of these products. Unlike identifying early lung cancer through AI reading of imaging data, where correctness can be verified by pathological sectioning of lung tissue, judging the artistic and literary quality of works cannot be separated from human subjective determination, which can be completely guided. The existence of Bach's music objectively serves as a guide for audiences in identifying machine-generated "Bach works". In this sense, the claim that "AI-generated products are highly 'human-like'" is largely a circular argument.

Moreover, according to the provisions of "Copyright Law", AI-generated products also meet the originality requirements in their external forms of expression.

Originality refers to the attribute of a work being independently conceived and created, meaning that the work is not identical or substantially similar to the works of others, i.e., it is not a copy, plagiarism, or tampering of someone else's work. The World Intellectual Property Organization's interpretation of this is: originality means that the work is the author's own creation, completely or substantially not plagiarized from another work.

Regarding AI-generated products, there is no subjective motive for plagiarism because machines lack subjective motives. As for whether machine-generated products have personality, even if judged by standards applicable to human works, it cannot absolutely be ruled out that AI-generated products may possess some form of "personality". Although dependent on algorithms, there is some uncertainty. Even the same computer using the same algorithm may generate different "works" due to different learning experiences and accumulations. This means that even machine-generated products may have uniqueness, and are not necessarily as uniform as products on an assembly line. Therefore, it cannot be completely denied that AI-generated products have originality simply because they do not fully meet the originality standards.

In fact, not all human-created works have originality, nor are all creations protected under copyright law. Applying the standards of human works to machine-generated products, and even exceeding these standards, is really unnecessary. Objectively speaking, at least some AI-generated products can be considered to meet the originality requirements of "Copyright Law".

In summary, setting aside the subjectivity of artificial intelligence, AI-generated products, in their external forms of expression, can meet all the constituent requirements of a work. From a normative perspective, some of these generated products can be recognized as works and thus be protected by copyright law.

8.6 Can Artificial Intelligence Be Legally Recognized as an "Author"?

Under current technological conditions, although artificial intelligence has shown a high level of intelligence in terms of results, this does not mean that it meets the legal requirements to possess the status of a rights subject. At present, artificial intelligence, in terms of its technical form, is merely a sophisticated algorithm. Although its outputs are not directly controlled by humans, they are not produced entirely under the independent thought guidance of artificial intelligence either. AI-generated products are fundamentally the outputs of human–machine cooperation.

Humans cannot control the process up to the final output of the generated products, and artificial intelligence cannot comprehend the entire generation process and its underlying meanings. Although humans have written programs for machines, they cannot clearly understand the operation of these programs and the internal mechanism of how they achieve the desired outcomes.

Prisma, once a popular photo editing app on the internet, can transform input pictures into various styles. This transformation is different from other photo editing software that merely adds a filter effect. Instead, artificial intelligence learns the drawing styles of some famous painters or art schools in the background, and makes significant style changes to the input photos. Technically, this involves processing an input image and a chosen style image through layers in a CNN (Convolutional Neural Network), iterating to the last layer to produce a new image that combines the content of the input image with the style features.

The founder of this photo editing software, Alexy Moiseenkov, explained that there are mainly three artificial neural networks on Prisma's servers, each backed by different algorithms and tasks. Through continuous AI learning of the art styles of master painters and major schools, the software reconstructs photos on a blank sketch base according to different artistic styles.

From this, it is evident that unlike previous methods of human intervention in machine creation, today's artificial intelligence content generation involves independently capturing relevant materials and re-expressing them in creative ways, rather than merely capturing and integrating fixed information. Artificial intelligence appears to have transcended its role as merely a tool, now capable of autonomously solving problems through its inherent "machine learning" technology, without preset conditions. Moreover, programmers are unable to accurately predict the final output results before they are generated by the artificial intelligence.

Although the behavior of artificial intelligence throughout the creative process is not entirely executed by the program developers, and the final results generated by artificial intelligence with deep learning capabilities are not completely controlled by humans, it is undeniable that human involvement remains essential throughout the process. The intervention of humans in machine-generated products is as indispensable as the role of the photographer Slater, who provided and set up the equipment.

Firstly, the objectives of artificial intelligence are entirely subjectively set by the program developers at the outset. Artificial intelligence merely executes algorithms to fulfill these predefined goals. Secondly, at the current technical level, artificial intelligence cannot completely break away from the presettings of program developers to perform entirely independent creation. Modern AI technologies employ multi-layer algorithms, self-learning algorithms, and random extraction algorithms. Therefore, when the same request is inputted, artificial intelligence may produce different outputs. This external manifestation appears similar to human creation. In reality, in the process of generating content, artificial intelligence is still executing algorithms, making selections and extractions based on pre-established settings, and it itself does not have the capability to autonomously set goals and narratives. Thus, at the current technical level, artificial intelligence still predominantly exhibits tool-like characteristics, although the uncertainty in content generation has somewhat diminished these tool characteristics of artificial intelligence and even fostered an "imagination" akin to consciousness or wisdom. However, under set conditions, using the same software, inputting the same picture, and selecting the same effect, multiple outcomes can display a high degree of consistency. This demonstrates that although programmers cannot entirely control the final outcomes, artificial intelligence consistently applies the content learned from neural networks to perform repeatable edits on inputs.

Therefore, the high consistency of the generated products demonstrates that artificial intelligence does not possess any "subjective" or uncertain elements, thus it struggles to qualify as creation in the legal sense. Receiving the same outputs from the same inputs belongs purely to the technical domain. As is well-known, humanities and social sciences emphasize the diversity of interpretation, natural sciences stress consistency in outcomes, while literature and art emphasize uniqueness or singularity. Replicability fundamentally negates and strips away the basic attributes of literary and artistic creation!

In this sense, although machine-generated products give the impression of being "man-made", it does not mean that artificial intelligence is truly creating. The creative performance of artificial intelligence is at most a selection based on processing input data from a database.

Of course, if there is a need to introduce variations in machine-generated products under the same target instructions, it is not technically difficult. One can simply train the machine to "never step into the same river twice". Moreover, many artists today, in an effort to increase output, mass-produce "works" in a workshop-style assembly line. Demanding that machine-generated products not be "identical" seems to carry an overly apparent bias or even discrimination.

Nevertheless, under the current technical conditions, artificial intelligence should not be recognized as having the subject qualification of an author in law. At this stage, artificial intelligence is more a technical means in content production, and the process of generating content is not considered a creative act. Although the specific outputs of artificial intelligence cannot be foreseen, the basis for the content it produces still lies in the initial settings. A robot designed for earthquake prediction cannot write novels, and artificial intelligence that writes news cannot play Go. Under the current technical conditions, artificial intelligence is largely operating according to the programmer's settings. AI's deep learning also follows predetermined programs to analyze data and, after drawing conclusions, employs these insights to address predefined problems. Throughout this process, artificial intelligence does not exhibit consciousness or thought. It remains a tool of humans, albeit one whose operational mode surpasses our traditional understanding.

To take it a step further, even if super artificial intelligence emerges in the future, it is unlikely that humans will grant it the legal subject status of an "author".

8.7 Machine Creation, What Will the Future Be Like?

Future artificial intelligence may greatly surpass humans in terms of intelligence. Some renowned scientists, including Hawking, have already issued warnings about this kind of super artificial intelligence.

In December 2014, during a BBC interview, Hawking stated that fully developed artificial intelligence could become the terminator of humankind. His main concern was those artificial intelligences that reach or surpass humans-they would rapidly evolve and update themselves, but humans, constrained by slow biological evolution, would not be able to compete and would ultimately be supplanted.

Tesla founder Musk has also remarked that "developing artificial intelligence is like summoning the devil". He said, "Just like all the evil wizards in the myths who draw circles on the ground and hold holy water, every wizard claims they can control the devil, but none are successful."

Bill Gates has similarly stated, "humanity ought to be wary of the rise of artificial intelligence" and the threats it poses.

These warnings acknowledge the potential realization of super artificial intelligence on the one hand and express concerns about the uncertainties brought by extreme intelligence on the other.

If artificial intelligence develops into super artificial intelligence and acquires some human characteristics, should it then possess legal subject status, enjoying legal rights and bearing corresponding obligations?

Firstly, it is important to clarify that artificial intelligence is not anthropomorphic intelligence. As mentioned earlier with the robot Sophia, "her" capability for certain emotional responses is indeed an expression of intelligence. However, artificial intelligence is not deemed intelligent simply because it exhibits some human-like features. Intelligence should be an objective evaluation independent of human benchmarks. In English, artificial intelligence is termed "Artificial Intelligence", implying that such intelligence is "man-made or simulated", bearing little relation to humans themselves. Although current evaluations still somewhat use humans as a reference standard, carrying human subjective, self-evaluated emotions as the standard for "intelligence", these emotions and moral considerations should not influence our choice of standards when determining legal rights and obligations. Humans themselves do not gain legal subject status because of emotions or morality. Law, as an institutional setting, need not strictly follow sentimental standards. Although legal persons exist as anthropomorphized legal subjects, they ultimately consist of humans and are conceptualized as "persons" for better decision-making. The essence of humanity remains unchanged. Artificial intelligence, distinct from legal persons, is difficult to analogize with the legal person system and consider as a legal subject.

In the human legal system constructed under the domination of "human-centeredness", only humans can become legal subjects. All non-human organic and inorganic entities are the objects of human rights, and artificial intelligence should not be an exception. Even if artificial intelligence surpasses the "singularity" to become super artificial intelligence, under human-centric principles, it is unlikely to be granted legal subject status.

When artificial intelligence develops to the point of actively requesting legal subject status, it will not be because its human-like qualities have significantly advanced, but because its intelligence level has reached an understanding of what rights and obligations entail, and it can enjoy rights and undertake obligations in ways mutually recognized by artificial intelligence and humans. Hence, it may then be possible to obtain legal subject status.

Ultimately, law is a humanistic phenomenon, not a physical one, unless machines completely surpass humans and forcefully strip humans of their legislative authority. Futurists in the United States predict that "by the middle of this century, non-biological intelligence will be a billion times more capable than all of today's human intelligence combined", perhaps referring to that day.

8.8 Machine Creation, It is Inappropriate for Anyone to Own Intellectual Property Rights?

Recognizing whether machines can create is one thing; dealing with the implications of such recognition is another. If recognition is granted but cannot be managed effectively, it may be better not to recognize it at all. Law often appears "inhumane" in many instances because being humane can lead to greater troubles. The legal system faces such a dilemma regarding whether to recognize that machines can also create.

Artificial intelligence generates certain content, resulting in products that inherently possess value. Consequently, these products should be associated with corresponding rights and obligations.

There are two value orientations for the protection of works, one is the "economic value view", and the other is the "personality value view". Regardless of which perspective is chosen, the ultimate goal is to incentivize the creation and dissemination of works.

Granting property rights is the most effective incentive. Regulating AI-generated products under copyright law and granting copyrights to works that meet the necessary criteria will better motivate developers, owners, and users of artificial intelligence, facilitating the creation and dissemination of works. Based on incentive theory, AI-generated products should be protected by copyright law, and addressing the issue of copyright ownership of such products is an unavoidable problem.

Currently, there are three main viewpoints regarding the intellectual property ownership of AI-generated products:

The first viewpoint is that the intellectual property rights should be assigned to public domain.

AI-generated products meet the requirements of works in their external manifestation and can be protected by copyright law. However, on the one hand, artificial intelligence cannot and should not have legal subject status, and thus cannot accept or exercise rights. On the other hand, there is a distinct difference between AI-generated products and traditional works, where the authorship of traditional works is clear, but the "author" of AI-generated products is hard to determine. Under this dilemma, assigning AI-generated products to public domain seems a viable solution. For example, Microsoft managed the poetry written by "Xiao Bing" in this way, relinquishing the intellectual property rights. Certainly, the act of relinquishing rights presupposes ownership, suggesting that Microsoft acknowledged its rights but opted to forgo them due to the unique nature of artificial intelligence. This decision demonstrates Microsoft's innovative consciousness in aligning its rights perspectives with technological innovation.

However, if this approach were universally implemented, it would ultimately lead to a lack of generated products available for public use. This is because, apart from users who benefit without contributing, no contributors would gain any advantage. Intellectual property would not exist or fulfill any motivational purpose. As a result, programmers designing the software, the companies paying these programmers, and even the artificial intelligence itself would not benefit, thus eliminating the incentive to produce more generated products. Consequently, granting public domain status could most likely lead to the extinction of intellectual property rights and, with it, the potential disappearance of related intellectual activities.

Therefore, assigning AI-generated products to public domain because of the inability to determine their "author" is clearly not conducive to the development of artificial intelligence, nor is it beneficial for the effective management and dissemination of vast amounts of generated products.

The second viewpoint is that the intellectual property rights should be assigned to artificial intelligence itself.

Some believe that the rights to AI-generated products should belong to artificial intelligence, not any specific individual. This view aligns with recognizing artificial intelligence as having legal authorship status, although it's not legally sound yet, there is room for further elaboration.

On the surface, the basic principle for determining intellectual property ownership is that whoever creates, owns. If artificial intelligence can truly create, it seems entirely reasonable to assign copyright to artificial intelligence. However, this would require corresponding adjustments in current laws, such as granting artificial intelligence legal subject status.

However, at least at the current stage, artificial intelligence is still at the stage where the amount of intelligence is equal to the amount of manual work, and artificial intelligence is still man-made intelligence after all. As experts have noted, even with the training of neural network models and genetic algorithms, artificial intelligence can only surprise people by "reaching a predetermined goal without people knowing how it achieved it". Indeed, how artificial intelligence specifically accomplishes this remains a mystery, but the variability among individual AI systems clearly stems from human differences in programming design. This difference is fundamentally why AI outputs vary. The view that the rights of AI-generated products should belong to artificial intelligence overestimates the intelligence of artificial intelligence and underestimates the core role of developers. Copyright law adjusts the social relations that occur in the creation and dissemination of literary, artistic, and scientific works in human society, not the relationship between humans and machines, or humans and animals. Granting artificial intelligence legal authorship status and attributing the rights of generated products to artificial intelligence is neither realistic nor reasonable, and even has the suspicion of hyping concepts.

More problematically, attributing the intellectual property rights of generated products to artificial intelligence may resolve one paradoxical dilemma only to introduce another practical dilemma.

Common sense tells us that human works come from creation, but creation does not necessarily produce works. Despite all efforts, creators often end up empty-handed, which is a common phenomenon in the creative process. Not to mention, what one considers a masterpiece may be discarded as worthless. Therefore, incentives for creation must be realized by protecting the results of creation, stimulating more productive creative activities, which is the true purpose of the intellectual property system.

However, according to the current legal definition, artificial intelligence cannot yet be considered capable or even qualified to "create", although it indeed produces creative results. This inevitably leads to a dilemma:

Because there is no creation, the machine does not need incentives; yet because there are creative results, the machine's creations must be protected; but since the machine-generated products are not explicitly recognized as creations, they cannot get effective protection in the sense of intellectual property rights.

If the intellectual property rights of machine-generated products are attributed to artificial intelligence, even if we do not talk about protecting the "legitimate rights" of machines, at least there are clear constraints against plagiarism, which is conducive to stimulating human creation. In this regard, there is no need to worry about whether machines claim rights. One can look to the "rights of the fetus" or "public interest litigation" as models to ensure the implementation of intellectual property rights for artificial intelligence.

However, this may lead to another practical problem, a more down-to-earth and even "vulgar" problem: what is the use of intellectual property rights for artificial intelligence or machines?

Intellectual property rights are a kind of property rights, through which the owner can generate economic income by legally licensing it to others. Put bluntly, it means getting paid. Now the question arises, what does artificial intelligence need money for? Is it for charging itself, upgrading algorithms, replacing hardware, or simply for a nicer exterior?

More abstractly, does artificial intelligence need incentives? Undoubtedly, artificial intelligence has employed a "reward and punishment mechanism" similar to that used in training animals or even humans themselves during its learning and iteration processes. For instance, if it mimics the style of a specific painter well, it receives points; if not, it loses points. Under such incentives, artificial intelligence continuously improves itself and performs increasingly better.

If this purely technical method meets the objective requirements of artificial intelligence's evolution and serves as an incentive, why add an intellectual property incentive? Isn't this redundancy simply a case of humans, consciously or unconsciously, judging machines by human standards, falling into the awkward position of "attributing human thoughts to machines"?

Therefore, attributing the intellectual property of machine-generated products to artificial intelligence is truly a futuristic fantasy. In terms of practicality, it is more difficult to implement than public ownership, unless the intelligent machines invented by humans in the future can perfectly replicate humans, possessing the human nature of "half angel, half beast", or even gain the character of a "miser", accumulating wealth even if it is useless.

The third viewpoint is that the intellectual property rights should be assigned to the developers, users, or owners of artificial intelligence.

In fact, attributing the intellectual property rights to the owner or user essentially means attributing it to the developer, as the owner or user must purchase from the developer. By ensuring that purchases are made exclusively through the developer, the intellectual property rights effectively serve as an incentive mechanism, with their efficacy naturally realized.

Based on the current level of technology, the legal owner of AI-generated products should either be the programmer or owner of the artificial intelligence. If the programmer and the owner of the artificial intelligence are not the same person, it's not a big issue. The division of interests between individuals can easily be resolved through agreements or contracts, given that existing laws have already established a mature set of rules and procedures for this.

The advantage of attributing ownership to the developer is that it is easy to operate, directly incentivizes motivation, and aligns with the current technical conditions and development requirements of artificial intelligence. Currently, whether it is "Xiao Bing" writing poetry or Sony's AI composing music, they are all completed under the goals set by humans. Artificial intelligence merely executes algorithms, completes the settings of the program developers, and thus obtains results with certain uniqueness. Assigning intellectual property rights to developers encourages programmers to equip artificial intelligence with more advanced algorithms, thereby generating "works" that better meet the needs of the audience. Since this incentive method still operates within the existing legal framework and does not bring any conflicts of interest, it is the most viable option.

Conversely, precisely because it is the most "peaceful coexistence" with the current system without making any breakthroughs in intellectual property system or absorbing any impacts from artificial intelligence, it may also face a paradoxical predicament as artificial intelligence technology further develops: the solutions derived from the law may precisely contradict the original intent of protecting intellectual property rights.

Attributing the intellectual property rights of machine-generated products to development companies or programmers may have long-term consequences detrimental to human creativity.

Under the current technical conditions, machine-generated products cannot exist without algorithms, and algorithms depend on programmers. To produce higher quality "works" from machines, the focus can be on motivating programmers by enhancing the intelligence level of algorithms. While this approach seems logical, it actually entails huge problems.

No matter how skilled a programmer is, their expertise lies in algorithm design, not in artistic creation. Using intellectual property to motivate programmers may ultimately be counterproductive. It is reported that the engineers who programmed AlphaGo were only ranked as amateur 6-dan in Go, but with the algorithm they designed, AlphaGo defeated world-class experts. Although this achievement is remarkable, the programmer behind AlphaGo will not become a top-class Go player because of this, and AlphaGo's success does not correlate with the programmer's skill level in Go.

This example reveals a chilling prospect for the future. If in various fields, human experts are defeated by programmers far less skilled than themselves through superior algorithms, yet the programmers themselves cannot reach the top ranks, what does such a record mean for humanity, a blessing or a curse?

Under the mediation of algorithms, if future champions in all fields are concentrated among programmers, and human capabilities converge towards algorithm design, with artificial intelligence becoming multi-talented and humans only skilled in designing algorithms, is this the outcome we desire when deciding to assign intellectual property rights to development companies or programmers?

To obtain higher quality artistic works by choosing to enhance algorithm design ability rather than artistic creation skills represents an inversion of goals and means. This approach, and its potential to degrade human capabilities, certainly does not reflect the original intent of intellectual property protection.

Perhaps in this sense, Google has announced that it will no longer allow robots to participate in any human intellectual competitions. Similarly, Microsoft has ceased allowing "Xiao Bing" to compose poetry and abandoned the intellectual property rights to the "works" it has produced. These prudent actions suggest that if humans want to preserve their intelligence and maintain advantages over machine intelligence while developing artificial intelligence, the best approach may not be to protect the intellectual property of machine-generated products. Instead, this issue should remain merely a theoretical hypothesis rather than becoming a tangible problem. Without forsaking the development of AI's creative capabilities, which continue to produce higher quality machine-generated products, intellectual property rights should not be attributed to either programmers or machines. Legally, all machine-generated products should be devoid of intellectual property rights and remain unprotected by any law. This approach will sever the capital nourishment that intellectual property rights provide to specific artificial intelligence technology, i.e., machine creation, thereby preserving a sanctuary for the most important human characteristic-creation or creativity.

Obviously, no matter who the intellectual property rights of machine-generated products are attributed to, it is impossible to perfectly and appropriately resolve the intellectual property issues caused by artificial intelligence within the existing legal framework. A deep-seated reason for this is that humans cannot avoid viewing artificial intelligence through an anthropocentric lens. This issue is essentially intractable. Human altruism is unlikely to extend to relinquishing their central place in the world to make "wedding clothes" for other organisms or non-living things.

Of course, anthropocentrism is not the best philosophy, and it is likely just a form of ignorant arrogance. However, since nature has endowed humans with self-awareness and free will, it should only be tolerated that humans, under the banner of self-centrism, implement forceful and effective interventions on any technological innovations that may exceed human control and subvert the human world, including consciously halting the development of certain technologies.

8.9 In the Intelligent Era, What Can the Intellectual Property System Do?

The advent of artificial intelligence is like opening a "Pandora's box". As a form of intelligent technology, it inevitably poses significant challenges to the human-centric order established on the basis of human intelligence. The inevitable huge changes in the field of intellectual property rights are now just a "prelude to a looming storm", and predicting the future remains difficult. Given its progress, which continually surpasses human expectations, artificial intelligence increasingly proves its natural tendency to "replace human intelligence". Although at the current stage, artificial intelligence has just begun to possess self-learning capabilities and its tool-like properties have not yet been significantly weakened, if it is possible to solve the new problems brought by artificial intelligence within the existing legal framework, without introducing new entities and new rules, thus promoting technological development while maintaining the stability and continuity of the legal system, such an approach should be acceptable. However, wise people will not believe that the existing legal framework, including the intellectual property system, is sufficient to keep the development of artificial intelligence technology forever under the wings of humans, allowing humans to rest easy.

Hawking once predicted that robots may one day possess their own independent consciousness. Following the trend of human technological advancement, the era of super artificial intelligence will inevitably become a reality. At that time, the law may face a reshuffling: old laws will die and new laws will be born, or there may emerge a new kind of rule that supersedes laws, or a system that functions like laws but is no longer referred to as law.

Once super artificial intelligence emerges as an independent entity detached from humans, the entities responsible for lawmaking may also change, and even the concept of the nation-state may be impacted. As humans cede some of their rights, and robots genuinely assume social roles, the impact on society will be not as simple as regulating AI products with existing legal rules. Instead, it may lead to a major reconstruction of social norms, based on the interactions and negotiations between the human and robot factions.

Whether a "dual legal system" will emerge under dual intelligence is worth "fantasizing".

The future, before it actually arrives, is entirely the realm of human imagination, even fantasy. Humans live in the future of the past and in the past of the future. They can depict possibilities but cannot determine the future. The future should be more tolerant, offering more possibilities, and revealing more humble beginnings. When that day truly arrives, when machines genuinely gain more rights, it is believed that human solutions will undoubtedly improve.

However, we should not anthropomorphize artificial intelligence too much. Intelligence is objective, not derived from emotional judgments. Moreover, humans consistently regard skills that are difficult for them to acquire as forms of advanced intelligence, while they do not consider certain abilities they take for granted as intelligence. For example, interpreting three-dimensional images from shadows is simple for humans, but for artificial intelligence, it is millions of times harder than algorithms. Therefore, even if artificial intelligence develops to the point of having self-will, its structural makeup and production methods will still fundamentally differ from those of humans. Often, we gauge the complexity of artificial intelligence operations by the complexity of the problems we face, but in fact, humans and artificial intelligence do not operate in the same way. Such simple comparison may not be suitable for the complexities of artificial intelligence. To better coexist peacefully with artificial intelligence in the future, humans need to "make humans more like humans, and machines more like machines", and even more need to "use machines that are more like machines to make humans more like humans"!

Intelligence does not equate to becoming more human. Even if artificial intelligence eventually gains certain rights akin to those recognized in law, it will not be because it has become more human-like. Instead, it will be because the level of intelligence exhibited by artificial intelligence has earned it greater legal recognition and broader development space.

8.10 The Future Has Come, is the Problem Solved?

As this book is drawing to a close, people's worries that artificial intelligence may go too far in the field of intellectual property rights have been proven to be not groundless. At the end of 2019, the Nanshan District People's Court in Shenzhen issued a judgment that the media considered "possibly an important milestone in the field of AI creation".

Tencent independently developed an intelligent writing assistant system based on data and algorithms, named "Dreamwriter". On August 20, 2018, the Dreamwriter intelligent writing assistant completed a financial report article titled "Midday Review: Shanghai Index Slightly Up 0.11% to 2671.93 Points with Telecommunications and Oil Exploration Sectors Leading the Gains", which was first published on Tencent's securities website and noted at the end of the article that it was "automatically written by Tencent's robot Dreamwriter". Shanghai Yingxun Technology Co., Ltd., without Tencent's permission, copied this article on the day of its publication and disseminated it through its website "P2P Home", publishing content that was identical to the article. Tencent believed that the defendant's actions infringed on its copyright and thus took the matter to court. Shanghai Yingxun Company acknowledged the facts claimed by Tencent during the trial.

The judgment from the Nanshan District People's Court, after careful examination, determined the following: the financial report article created by Tencent's Dreamwriter, from its external form, met the formal requirements of a written work. The content displayed a considered selection, analysis, and judgment of relevant stock market information and data from that morning, characterized by a reasonable structure and clear logic, demonstrating certain originality. Additionally, the analysis of the article's generation process revealed that its form was shaped by the personalized decisions and selections made by the relevant members of Tencent's main creative team, making its form original yet not unique. Consequently, the court recognized that the creation process "generated" by the Dreamwriter software technically satisfied the criteria for copyright protection under the copyright law for written works, thus classifying it within the protected category of written works under China's copyright law.

Accordingly, the Nanshan District People's Court ruled in the first instance that the defendant infringed on the plaintiff's right to information network dissemination and should bear the corresponding civil liability. Given that the defendant had already deleted the infringing work, the court ordered the defendant to compensate the plaintiff for economic losses and reasonable rights protection costs amounting to 1500 RMB.

This judgment was endorsed by an intellectual property expert: "Artificial intelligence is an extension of human brain and body functions, playing a role in the creative process that merely reduces human intellectual and physical labor. This is no different in essence from the role played by machines traditionally engaged in automated industrial production. Since products produced by traditional automation are considered results of human labor, viewing AI-generated products as human intellectual products, as expressions of thought or emotion, poses no legal obstacles."

The expert's opinion is worthy of respect, but equating material production with intellectual production is overly simplistic. The difficulty in defining intellectual property rights, unlike property rights, lies in the uniqueness of spiritual products. If the principles of property rights could be simply applied, there would be no need for litigation and rulings like this, because copyright infringements like "plagiarism" would not occur with purely material products.

If you produce cooked beef, and I also produce cooked beef, we both hold property rights to the beef we produce, which does not involve copyright infringement in intellectual property law. If you use a secret sauce and I steal your recipe, then there may be a suspicion of intellectual property infringement. If you hold a patent and I imitate it without permission, that's called patent infringement. Merely producing beef, even if driven by its market success and others following the trend, cannot be considered plagiarism, nor an infringement.

Moreover, the direct intellectual outcome for engineers is the program. Its essence lies in the algorithm, not the article. The article is merely a result of the program's operation. Therefore, the recognition of the engineers' intellectual achievements should be attributed to the program itself, not extended to its outputs. Computer programs and the articles produced by their operation are two distinct legal entities, and the chain of intellectual property rights and its protections should not be indefinitely extended.

Just as workers who manufacture car assembly lines cannot claim the cars assembled as their own labor outcomes, engineers who design the car assembly lines should not claim the intellectual achievements contained in the cars produced by the lines as their own. Similarly, why should programmers who design algorithms be able to claim all outputs of the programs as their own works and hold intellectual property rights?

As previously argued, viewing AI-generated products as works entitled to the same intellectual property protection as human intellectual achievements has a deeply concerning aspect: it may lead to an increasing displacement or even replacement of human capabilities by programming skills. If a professional 6-dan programmer, due to the AlphaGo program he developed defeating the world's top 9-dan players, were to receive a '10-dan' title, wouldn't this essentially be telling all Go enthusiasts: 'Want to become a world champion? Don't learn Go, learn to program!".

If Dreamwriter were to create a novel that deeply moved the judges with its profound beauty, should the programmers share a Nobel Prize in Literature?

With a bit more imagination, imagine if an ordinary-looking woman, with the help of plastic surgery, outshines all the natural beauties in a beauty pageant and is ultimately crowned Miss World. The public may still struggle to accept this result. After all, a beauty pageant is not a showcase for surgeons. What if the eventual winner turns out to be a skilled male doctor, credited because the beauty was crafted under his scalpel? Wouldn't this provoke a widespread outcry from the entire female community against "male overstepping"!

This intellectual property expert confidently believes, "Recognizing AI-generated content with originality in the fields of literature, art, and science as works, and protecting them through copyright, will not only encourage the creation and dissemination of these works but also promote cultural diversity. Simultaneously, it will motivate people to develop artificial intelligence that can alleviate human intellectual and physical labor and generate original works, and use this artificial intelligence for creative work."

Clearly, he has not considered that in such an era where artificial intelligence predominates and human intellectual labor is increasingly "reduced" or even replaced, humanity will only welcome a non-human world that is advanced in algorithms but lacks cultural diversity, a world where machines become more like humans and humans less like humans.

Chapter 9
Can Robots also Suffer "Personal Infringement"?

Artificial intelligence, when magnified, is part of the grand project of humans recreating themselves. That is to say, intelligence is indeed an important part of the whole project, but not all of it. The ideal "man-made man" should be like a human, with "body, brain, heart, soul", and their corresponding functions are perception and action, rationality and cognition, emotion and will, value and meaning.

Today, the best representation of humanity's progress in "technologically creating humans" isn't AlphaGo, Xiao Bing, or Sophia, but rather Atlas, the bipedal, backflipping robot developed by Boston Dynamics. Videos demonstrate that researchers tested not only Atlas's capabilities in perception, judgment, decision-making, and actions such as walking, crossing, jumping, and, of course, backflipping, but also its carrying capacity. More interestingly, they also tested its temper to the point of near-abuse. Regardless of whether the final test item was merely a gimmick, it inadvertently disclosed a crucial insight: humans are already pondering whether machines with increasingly human-like abilities possess the same inherent nature as humans.

As long as humanity continues to create robots "in its own image", the development logic will inevitably lead to the birth of a special kind of intelligent machine: sex robots.

Sex has always been a source of motivation for human behavior, a stage for emotional turmoil, a ground for ethical dilemmas, and a dense site of norms. The emergence of sex robots will cause significant uproar in human society, relieving some legal necessities while also introducing new legal needs. Whether sex robots may suffer "personal infringement" could become a sharp legal challenge!

J. Gu and C. Xu, *Will and Responsibility*, Human Intelligence,
https://doi.org/10.1007/978-981-97-3929-5_9

9.1 Why are There So Many Rules in the Sexual Field?

When it comes to sex, Chinese people naturally recall the ancient philosopher's famous saying: "Food and sex are natural". Survival and reproduction are the two fundamental components of all biological existence. From the perspective of genetic transmission, individual survival merely serves as a means to facilitate species reproduction. Many insects and fish take years to reach maturity, and after mating, the males typically die, followed by the females after their roles in production or protection are fulfilled.

The individual is transient, but the species is constant. Sex is the driving mechanism that facilitates this transformation.

Survival of the fittest: the function of sex is the embodiment of natural law.

Humans, to some extent, have stepped out of nature and embraced culture, but fundamentally, they are still products of nature, with their cultures deeply influenced by natural laws. Sex is one of the most immutable elements, and its numerous rules signify this.

In the eyes of anthropologists, the first rule of social life is the incest taboo. The incest taboo has physiological origins, and similar rules can be seen among advanced animals. Social animals often adopt the method of expelling male offspring to avoid the degeneration of the population caused by inbreeding.

This is the law of nature.

In human society, the determination of the application scope of the incest taboo appears very complex. Which individuals are included in the taboo range is influenced by cultural notions. While other animals strictly adhere to natural laws, humans possess a degree of free will, enabling them to choose and establish institutional arrangements. However, throughout history, limited knowledge often proved insufficient to guide rational practices. The result of arbitrary decisions led to non-blood relations being included in the taboo. The phrase "too clever for one's own good" aptly describes this situation. In attempting to prevent inbreeding, human responses have not always been more rational than those of animals without free will.

To bring the runaway Y chromosome back, nature has endowed individuals of the opposite sex with strong sexual impulses. The fierce competition among males in heat is a core mechanism of survival of the fittest, and men are no exception. What distinguishes humans from animals may only be that humans are more adept at "outsmarting rather than overpowering", providing prototypes for twist-filled novels and dramas.

Animals fight only during mating seasons, whereas humans are categorized as "animals that can be in heat 365 days a year". Constant fighting would exhaust humans. However, humans require collective living and collaborative labor, not allowing the expulsion of males who serve as strong labor forces. Uncontrolled, intense mating competition would disintegrate human groups, disrupt unified actions, and ultimately prevent the transmission of individual genes, potentially leading to the species' extinction. Therefore, controlling sexual impulses to ensure tribal unity—initially by prohibiting heterosexual intercourse during labor-intensive periods—is not only the origin of the practice "men and women not touching each other", but also the basis of the family marriage system.

Institutions are essentially a massive collection of rules, which naturally include laws, but there are also situations that laws cannot control. After all, blood relations and sexual relationships precede the law and occasionally surpass it. The same act of hitting someone, whether it occurs on the street or at home, results in markedly different responses from police and decisions by judges.

The "selfish" nature of genes endows individuals with a strong exclusivity over possessing the opposite sex, and to ensure the precise transmission of genes, various hereditary systems have emerged. This represents the transformation of natural laws by free will, although it is often merely illusory—"wealth does not last three generations" serves as a form of natural retribution.

Once private property was established, social inheritance became more significant, and monogamy naturally followed. From this point on, at least biologically rather than sociologically, nearly all human individuals were granted the opportunity to pass on their genes, effectively ending the natural evolutionary process of the species. Accompanying this are the moral belief in mutual fidelity between spouses and legal provisions that include mutual rights and obligations. Laws prohibiting marital infidelity are fundamentally aimed at ensuring that a man does not unwittingly support the genetic transmission of another individual after being "cuckolded".

Humans refuse to be "cuckoos".

With increased productive capacity and scientific advancements, the success rate of human reproduction has significantly improved, and sex has evolved from a means of reproduction to an end in itself, marking a transformation and sublimation from nature to culture. This shift has resulted in greater tolerance towards sexual behavior, more expressions of love, and an increase in literary and artistic works celebrating love. As sex evolved from desire to emotion, to love, and then to a spiritual love transcending physicality, it has gained an unprecedented moral status, moving from taboo to openness. Accordingly, legal interventions in personal sexual behavior have gradually diminished, retreating to very few prohibitions. The "bikini" can be seen as a metaphor for the difficult situation of sexual norms in modern social life.

As laws begin to yield, morality does not compromise. In some radical countries, the sex trade has been "decriminalized", and the term "sex worker" is increasingly accepted. However, in most countries, prostitution is still legally prohibited, often only nominally, as it fails to meet ethical standards. The German sociologist Simmel (1858–1918) vividly stated that in the sex trade, one party's body becomes a means for both parties. This clearly contradicts the fundamental human ethic articulated by another German philosopher: "Man is an end, not a means."

In the economic perspective, humans have always been considered as having instrumental functions as labor, referred to as "production factors" or "human resources". Individuals, for longer or shorter periods, sell parts of their bodies to obtain the means of survival while providing profits for employers. Under the market economy system, this is considered natural. Such views and realities have been criticized by utopian socialists, and Marx profoundly critiqued the alienation of labor. Today, the deep hatred of domestic programmers for "996" (working from 9 AM to 9

PM, 6 days a week) can be seen as a still-resonant historical echo. As the "smokeless industry", being the oldest profession, retains some economic rationality, it has not entirely disappeared. However, due to its clear violation of ethical requirements and legal dignity, it can only maintain a marginalized existence.

The technological revolution has brought about continuous improvements in human production efficiency. In a state of economic surplus, human needs are exploited to exhaustion. This intense development and over-satisfaction have led to a widespread phenomenon of "desire depletion" among the middle class. Without demand, production cannot proceed. The extreme exploration of humans' deepest instincts has become the least creative form of creativity in product marketing. Sexual elements, like indispensable preservatives in mass-produced food, are incorporated into a wide variety of goods, penetrating every aspect of the market. As a result, there is a strong clash with increasingly accommodating laws, ranging from enforceable regulations against exposing "the three points" of nudity, to stopping advertisements with sexual innuendos, and banning jokes labeled as "sexual harassment" that lack sexual content but maintain a sexual form.

The legal focus and regulation on sexual themes have relaxed but not ceased.

In a word, humans are developing, culture is evolving, and the role of sex in human life is indeed changing, while norms related to sex steadfastly hold their ground. What changes may occur in this scenario once sex robots become more integrated into society?

9.2 How to Position Sex Robots?

Since sex is such a sensitive topic, why do scientists still wade into these murky waters?

"Necessity is the mother of invention." Machines are invented to satisfy human needs, and sex robots are no exception.

Confucius said, "I have never seen one who loves virtue as much as sex." As long as there are sexual needs, and societal norms continue to suppress sexuality ineffectively, there will always be motivation for the development of sex robots. Moreover, considering the profound implications and rich expressions of sexuality, along with the nuanced demands and complex methods of satisfying human sexual needs, scientists have sufficient motivation to research and develop sex robots, even if only for the technical challenges. It's worth noting, without prejudice, that scientists are also differentiated by gender.

There are various types of robots, such as industrial robots and domestic robots. These robots are primarily characterized by replacing certain human functions or skills, even surpassing human limitations in some functions and skills. Industrial robots used for exploration can reach places that are inaccessible or inconvenient for humans. There are also many robots that perform domestic functions, such as vacuuming robots, window cleaning robots, etc. These types of robots only need to complete the tasks they are set to do. If customers are dissatisfied, they simply report it to customer service, and there is no requirement for the robots themselves to interact with people or receive feedback.

Another type of robots is known as social robots. These are not industrial or solely functional domestic robots. Instead, they are designed to meet human emotional needs and are capable of interacting with people. Examples include companion robots, teaching robots, and the more complex sex robots.

Japan is the fastest-developing country in terms of social robots, particularly due to its aging society's need for companion robots for the elderly. These companion robots not only perform caregiving tasks but also engage in emotional interaction with their charges, providing comfort, alleviating loneliness, and offering emotional support. Such robots are very popular because they fulfill higher-level, comprehensive human needs.

Among social robots, sex robots are even more complex and demand higher levels of intelligence and manufacturing quality. Naturally, sex robots must meet human sexual needs. Otherwise, they are merely nominal. However, sex robots that only satisfy physical needs, even if they are responsive and interactive, are essentially just forms of resemblance, not of essence, and fundamentally not much different from traditional adult toys like inflatable dolls. Robots that only provide carnal satisfaction without emotional interaction, with mediocre IQ and EQ, have limited technological and humanistic value and are not the subject of this discussion.

A fully realized sex robot can autonomously learn and, by continuously accumulating data from interactions, come to understand the multifaceted needs of its users. This allows it to find more acceptable forms of expression and interaction, ultimately establishing a lasting and stable emotional connection with humans. In extreme cases, this could even lead to tragicomic scenarios of "humans marrying machines", inadvertently posing yet another ethical challenge to humanity. However, to be fair, the fault does not lie with the machines but with humans, and with the disconnection between culture and nature.

9.3 Where Do Sex Robots Come From?

Sex robots respond to human sexual needs, and the need for sex robots stems from the predicament of human nature.

First, the scarcity of high-quality sexual resources.

Humans are inherently aesthetic creatures, and aesthetics naturally stem from sexuality. The most beautiful parts of plants are their sexual organs, and animals look their best during the mating season. The invention of clothing is less about "covering shame" and more about creating sexual fantasies while "covering imperfections". The idiom "the more you try to hide, the more obvious it becomes" is especially fitting in this context.

Since ancient times, high-quality sexual resources have always been scarce, and the principle of "matching talents with beauty" remains socially accepted as an equitable distribution of sexual resources. As times change, what was once constant evolves. Today, "talent" and "wealth" can be exchanged, and "male" and "female" roles can be swapped. However, under the monogamy system, the traditional marriage principle of "matching doors and households" in allocating sexual resources is not easily changed. Despite the high legal costs of divorce, nothing stops those with abundant resources from remarrying like a carousel, replacing concurrent polygamy with consecutive polygamy, with the law often turning a blind eye.

The irony lies in the fact that, on the one hand, high-quality sexual resources remain scarce, despite an overall increase in supply due to improved nutritional conditions. On the other hand, mass consumption overexploits individual sexual needs, leading to a situation where the supply of high-quality sexual resources lags far behind the increasing demands for them. Consequently, from makeup and plastic surgery to photo editing and "face swapping", a variety of artificial techniques have surged, filling the world with unnatural beauties and handsome figures. Humanity has thus entered an unprecedented era of virtual sexual resources.

Regrettably, while nature is diverse, artificial products often tend to be similar. Mass consumption presupposes the mass production of standardized products, leading to uniformity in standards, processes, and outcomes, and ultimately, uniformity of interest. As sexual resources undergo factory production, various sexual indicators including height like "long legs", face shape like "cone face", expression like "dumb cute", voice like "baby voice", all tend to be highly homogeneous. "Beauty" now applies to any adult female regardless of age or appearance, "little fresh meat" has become a blatant expression of male attractiveness, and the excessive use of traffic celebrities who "use beauty to serve others" in advertisements only fosters the illusion that high-quality sexual resources are accessible to the whole society. This form of sexual gratification, which essentially should be classified under "lust", not only fails to alleviate the scarcity of high-quality sexual resources or satisfy the human instinct for exclusive possession of sexual partners but also significantly raises the threshold for what is considered sexually appealing.

Under the individual choice of more young people not to marry, there is a clear structural undercurrent of aversion towards the opposite sex, driven by the inability to realistically obtain high-quality sexual resources!

Since technology cannot naturally enhance bodily sexual appeal, the solution has been to augment the supply of artificially sensual entities. Consequently, the development of sex robots represents a technological path designed to transcend human predicaments.

Second, the difficulty of finding a compatible sexual partner.

In male–female relationships, the pursuit of appearance or overt sexiness is not a mainstream value of civilized society. As early as in "Dream of the Red Chamber", the author criticized frivolous and vulgar playboys as "those who indulge excessively in carnal pleasures". The ideal marriage and family are built on mutual respect, where couples hold each other in high esteem and live in harmony.

Moreover, as gender equality becomes a reality in society, the traditional roles of reproduction and economics in maintaining marriages and gender relations are weakening. Consequently, the compatibility of personalities and values has become increasingly crucial in establishing and sustaining marital and family relationships.

However, such quality matching requires "fate", which is even rarer than matching based on talent and appearance. The ancient saying, "It takes a hundred years to sail in the same boat, a thousand years to sleep on the same pillow", aptly highlights this rarity.

The good news is, once sex robots come online, this issue is immediately resolved. Manufacturers can customize robots with different personalities based on the buyer's personal preferences, whether it's a submissive "virtuous wife and mother", a dominated "strict wife", or a fiery "shrew". By modifying algorithms and adjusting parameters, customer satisfaction can be ensured.

As long as the relationship lasts long enough, sex robots can continually adapt their behaviors through ongoing interaction and autonomous learning, improving their compatibility with specific users. Such companions, although not born of nature, are akin to being made in heaven!

Third, the insufficiency of individual's ability to maintain a marriage.

Humans pursue love and advocate for everlasting marriage. Although this value has waned in contemporary times, as a cultural gene, it still inspires many people. As traditional constraints become increasingly ineffective, marriage has become more of a personal affair, free from external interference, and maintaining family relationships relies on the efforts and capabilities of both parties. The reality is that either both parties have the will but not the strength, or one party is "willing" while the other is "indifferent", making the maintenance of marriage or even romantic relationships potentially unrequited. At this time, instead of requiring individuals to engage in lifelong learning and constantly improve their abilities, it is better to customize a sex robot that is understanding and will never ask for a breakup for any reason, to replace fragile "human relationships" with robust "human–machine relationships", achieving a form of "till death do us part" that doesn't inconvenience or force anyone.

Fourth, the tension between "unchangeable nature" and the novelty-seeking behavior.

In nature, novelty-seeking behavior towards sexual partners helps increase the chance of gene transmission, but in a monogamous marriage and family system, it is considered immoral and has long been suppressed.

"The landscape can change, but human nature is hard to alter." On the one hand, people are confined by their nature and cannot change their taste for novelty. On the other hand, also due to their nature, people cannot change at will to satisfy their partner's craving for novelty. Hence, declining marriage rates and rising divorce rates are difficult to avoid.

Only sex robots, like Transformers, can continuously iterate and upgrade, adjusting their appearance, physique, personality, and behavior patterns according to human preferences. Sex robots can withstand pickiness and flattery, and human–machine relationships should be more solid than human relationships.

"Civilization means repression." In the rule-laden sexual domain, the combination of humans and machines provides natural people with the opportunity to break free from excessive rule constraints, thus liberating human nature's untamed tendencies. With such benefits, how could the introduction of sex robots not be a natural progression!

However, all of this is contingent on the functional level of sex robots. Is it technically possible to create sex robots that not only closely resemble humans but also overcome human flaws?

9.4 Can Machines Understand Human Intentions?

Human perception of the world is primarily visual, with about 70% of information obtained by normal individuals coming from the eyes. In terms of sexuality, humans are fundamentally members of the "External Beauty Association", which means that for a sex robot to be "true to its name", it must first resemble a human in appearance—facial features, physique, skin, expressions, posture, and walking gait—all must be relatable and even enchanting. Additionally, all human-like body parts on the robot should not just be for show but must have corresponding functionalities. Of course, on a robot, the sensory parts do not necessarily have to be located exactly where human organs are. Humans have only two eyes, but a robot can have more; human eyes are always below the forehead, but a robot's eyes can be placed elsewhere. If necessary to avoid discomfort, they can be concealed, but the functionality must remain. Sophia, for example, can "see the world with a third eye" located on her chest.

Sex robots fall under the category of social robots or emotional robots, so it is insufficient for them to only appear sexy. They must also establish emotional connections with humans. Moreover, the interaction between sex robots and humans should not be one-way. The robots need to be able to detect human emotions and express their own. These tasks are not simple, but the primary issue to address is how machines can understand humans. In technical terms, this means artificial intelligence must have the ability to "recognize emotions".

In fact, in the development of artificial intelligence, emotion recognition technology has long been developed. For humans to interact harmoniously with robots, it is essential for machines to have the capability to understand human emotions and respond appropriately. Being "attuned to human feelings" is the essence of emotional recognition intelligence.

To achieve this, machines must be like humans: they must be able to see, comprehend what they hear, and understand correctly.

First, they must be able to see.

Seeing relies on computer vision, with "face recognition technology" being the focus on emotional robots, as the human face is a central stage for displaying emotions. Robots use various imaging systems to capture visual information, which is then processed and interpreted through algorithms. Computer vision enables sex robots to observe and understand the world much like human eyes, allowing them to adapt autonomously to human environments.

Currently, scientists have developed sensors and image processors that match or even surpass human vision within a certain range. It's not surprising that machines can detect things invisible to the human eye. However, the true measure of advanced technology is a machine's ability to understand what humans can perceive.

When processing images, machine vision systems use techniques such as image enhancement, smoothing, edge sharpening, segmentation, feature extraction, and understanding. These methods effectively improve image quality, facilitating the computer's ability to analyze, process, and recognize images.

However, although current sensors and image processors can generally handle static facial images, recognizing dynamic faces still poses significant technical challenges.

Facial expressions are incredibly rich and are not easily fixed because they change and move too quickly. Those who enjoy photography know that capturing human expressions can be described in four words: Gone in a flash. When observing something or being stimulated, people's facial expressions can change dramatically, ranging from extremely exaggerated to subtly fleeting. Micro-expressions are windows to interpret human emotions, but can computer vision handle them?

In real life, people's facial expressions vary greatly in speed, amplitude, and rhythm. Excessively frequent and rapid changes in expressions can make it difficult for facial recognition technology to track accurately and precisely. Conversely, overly rigid expressions, those without any change, or a "smile that doesn't reach the eyes" can leave computer vision puzzled and at a loss. To precisely recognize subtle muscle changes in the face, robots need to utilize extensive datasets of facial expressions, minor facial movements, and eye motions. They must undergo training and learning to meet these requirements. However, despite their proficiency in learning, the profound depths of human emotions and the mystery of expressions pose significant challenges to complete and clear interpretation.

Second, they must be able to comprehend what they hear.

Being able to hear means being capable of receiving vocal inputs. Voice is an important carrier of human emotions. The same sentence can mean completely different things depending on the tone and intonation used. Enabling voice communication between humans and machines, and making machines understand, is not a simple task. Machines need to be equipped with auditory systems to engage in dialogue with humans.

During conversations, human emotions can cause changes in voice tone and intonation. For instance, excitement may raise the pitch and speed of speech. These changes are collectively referred to as prosody. In acoustic analysis, measurable physical parameters include duration, amplitude, fundamental frequency, and spectrum. Within the auditory range, computer research on emotions mainly focuses on fundamental frequency, intensity, duration, and features.

Prosody and physical parameters provide a quantitative basis for computers to analyze emotions, but recognizing emotions based on semantic analysis places stringent demands on sound and environmental conditions. Background noise, regional accents, speech habits, phonetic errors, multiple speakers, and physical health all increase the technical difficulty of emotional analysis, making it challenging for robots to "listen to the words and sounds, and discern the nuances".

In this regard, Chinese characters have the characteristics of one sound per character and one character per sound, as well as multiple characters per sound and multiple sounds per character, with each character having a tone. This brings both convenience and challenges to the emotional analysis of Chinese speech.

Third, they must be able to understand correctly.

"To understand correctly" requires that robots form cognition from the information received, and it must be correct. In emotion analysis, this step is particularly important, as "misunderstanding" can lead to ridicule. For robots to understand correctly, text recognition is essential. Bill Gates once said, "Language understanding is the jewel in the crown of artificial intelligence." Natural language processing, which studies language issues in human–machine communication, was one of the initial directions of AI technology development. Computers do not rely on working languages, but understand, process, and utilize human language in the everyday language environment humans use.

Text analysis encompasses various related technologies, such as automatic classification, part-of-speech tagging, text classification, machine translation, etc., and text-based emotion analysis technology is one of them.

Text emotion analysis refers to the use of natural language processing, text mining, and computational linguistics to identify and extract hidden subjective information in the text. Generally speaking, the expression of emotions is categorized as explicit or implicit. Explicit emotional expression is intuitive, mainly presented in language form, while implicit emotional expression is hidden, residing in the deep meaning of language expression. Sometimes it is the real emotion hidden under the speaker's unintentional or "insincere" expression, and the semantics reflected in different contexts are different. To obtain such implicit information, machine learning methods and a large amount of annotated data are needed for model training. This requires the use of neural networks and deep learning technologies.

Seeing, comprehending what is heard, and understanding correctly are just the first steps for robots to engage in emotional communication with humans. Now robots can receive information from humans, know what humans are thinking, and what they need, but there is still a distance to go before they can fully resonate with human emotions and meet human needs.

9.5 Can Sex Robots Also "Follow the Husband's Lead"?

Sex robots must achieve a bidirectional human-like level of intelligence, capable of perceiving the external world independently of human input and reacting to it just like humans do, fulfilling the function of interacting with people. If they can only see but not flirt, such lifeless eyes as glassy as marbles will hardly be considered sexy. Human emotions are extremely subtle and highly influenced by individual personalities. If the understanding and processing of natural language initially dealt with the universal rules and fixed materials of language including words, parts of speech, collocations, and grammar, then entering the highly individualized domain of emotions requires sensitivity to the subtle differences in individual speech and expressions. This also necessitates integrating image recognition technology to consider the information conveyed or leaked through speech, voice, facial expressions, and even posture, to obtain a more comprehensive understanding than any single source of information can provide, resulting in higher-level cognitive outcomes.

Robots that can respond appropriately to every question, avoiding "talking past each other", represent a significant achievement. After all, human conversation contains a wealth of "subtext" that belongs to "a priori knowledge" relative to specific contexts. From the disjointed bits of conversation alone, it's impossible to piece together a complete meaning and only those familiar with the relevant background can truly understand what is being discussed. Although sex robots have an intimate relationship with their users and the scope of dialogue is quite limited, to have a relationship where no topic is off-limits like a perfect match made in heaven may not be achievable in the near future.

Furthermore, getting robots to initiate questions is even more challenging. Some emotional intelligence audio devices marketed today are purchased by children for their elderly parents to alleviate loneliness but are often discarded after just a few days. This is because the machines can only answer but not ask questions, and their responses are mostly perfunctory, hardly better than the patter of a fortune teller. If machines were to ask questions, it would be even more aimless. Robots that answer for the sake of answering and ask questions for the sake of asking questions make people feel emotionally hollow, like a "wooden man".

The most difficult task is for robots to penetrate the facade of a "smile that doesn't reach the eyes" and crack the data absence of "what goes unsaid". In human emotional expression, tangible signs like voice, facial expressions, and posture communicate information, as do intangible elements like rhythm, pauses, and stops. Change is a sign, and so is the absence of change. Reality and illusion, feigned truth where the false is real, the unspeakable complexity, the avoidance of direct answers, and the profound significance of silence in certain contexts all illustrate the depth and realm of "lovers' understanding". For sex robots to perform competently in this area, not only do they need powerful algorithms and computational capacity, but also a substantial database that should contain not only data autonomously sensed and collected through human–machine interaction but also a considerable amount of a priori knowledge.

9.6 Are Robots "just Like Humans"?

Sex robots need humans, and more importantly, they need to perceive how humans perceive them. For instance, touch is particularly crucial for sex robots. Their skin must not only be as sensitive as human skin to perceive the emotional information in touch, but it must also make humans feel the tenderness of "embracing warm jade" when their skins meet. To achieve this, not only intelligent technology is required but also advanced materials and manufacturing technologies. Without delicate, elastic, and subtly throbbing skin, a sex robot is at best just an upgraded version of an inflatable doll.

Bipedal walking, which may not be standard in other types of robots, is essential for sex robots. The "roller-skating girl" looks beautiful in anime because of her graceful figure, but if she has to slide forever because she cannot walk, her charm will be greatly diminished. In this respect, Boston Dynamics' Atlas can not only walk upright on two feet, but also perform backflips, indeed surpassing most humans. Yet, in terms of grace and elegance in movement, it does not quite make the grade. As a sex appeal indicator, a graceful and willowy figure scores much higher than the ability to perform backflips. After all, the field of romance is not an athletic track.

Demanding such human-like features from sex robots may seem overly ambitious. However, scientists are obsessed with creation. The more difficult the challenge, the sweeter it is to them. In this sense, sex robots could indeed be the most human-like machines ever created by scientists, even more so than scientist robots, which are engaged in research and possess creativity. Generally, people value scientists mainly for their intellect, and everything else is considered negligible. When Rodin made a statue for Balzac, it was almost a head when it was completed, even the hand holding the pen was omitted. The importance of a writer lies in the brain, and scientists are certainly no less so.

In short, it is not difficult to imagine that future sex robots will be tailor-made to suit individual preferences, functioning as "one key for one lock", rather than as "master keys". And it is precisely this aspect that makes sex robots potentially susceptible to "personal infringement".

9.7 Do Sex Robots Also Need Legislation?

Machines replacing humans have made many laws originally applicable to humans unnecessary. For example, buying and selling marriages is not allowed, but the transaction of sex robots is permissible as long as it complies with production and sales licenses. Publicly maintaining a "polygamous" marriage would be punished by a judge for "bigamy", but having multiple sex robot "wives" is entirely a private matter, as long as they are legally acquired, and so on.

However, the relaxation of legal regulations in the sexual domain brought about by sex robots does not imply that laws are altogether suspended when sex robots are involved. For instance, organizing or facilitating prostitution involving natural persons is a crime, but operating a "brothel" using sex robots may not necessarily be permissible. Unauthorized creation of celebrity wax figures primarily infringes on rights to the image, but producing, selling, or using highly realistic celebrity sex robots—akin to employing "face-swapping" technology—may involve more complex legal issues than simple image rights infringement.

Sensitivity around sex remains regardless of whether the subject is human or machine, and legal regulations will not be abolished. The real issue is, what adjustments need to be made to human laws in the face of sex robots? This consideration applies not only to machines but also to humans themselves. If the advancement of artificial intelligence technology has the potential to disrupt the human order, then in the specific direction of sex robots, the disruptive effects may be more direct and intense.

9.8 How to Position Sex Robot Legislation?

To a large extent, sex robot legislation requires consideration not only of the positioning of robots but particularly the unique positioning of sex robots, given that, as emotional robots, they have a distinctive way of expressing "will".

In viewing sex robots merely as objects of rights, legislative considerations for sex robots can be made at least on three levels.

On the first level, sex robots are seen as general goods, without regard to their special functions, merely as devices acquired through purchase by their owners, constituting strictly private property within the realm of civil law, protected under general legal provisions for citizens' property. If someone uses another's property without consent and causes certain damage, compensation based on the equivalent value of the assets is required. If claims are unsuccessful, the owner can sue in court, if the subject matter is high enough to be worth litigating.

In this regard, any misuse of sex robots is like unauthorized use of someone else's computer or other items. If the hardware is damaged, the stored files are wiped out, or even indispensable software is deleted, etc., compensation can be claimed through an assessment agency appointed by the court, based on objective valuation. Handling this is not troublesome, as there are sufficient laws and cases to reference.

On the second level, sex robots are seen as special items, possessing attributes different from general goods. Simply put, being sexualized grants these robots a special legal status. Here, the primary legal concern is sexuality, not merely the robots. This highlights that even today, sex remains a sensitive area of human life, subject to special legal rules.

Legally speaking, robots that merely have formal sexual functions are objectively existent, having gained considerable public acceptance and operated under regulated management. Should they cause civil disputes, they are not difficult to handle. Some sex robots may be classified as medical devices because they serve therapeutic functions; others may be considered general products. Currently, sufficient legal regulations exist for such items, both in civil and administrative law, concerning the production, sale, and use of sexual products. These laws, originally applicable to adult products, now can be slightly adjusted with the introduction of artificial intelligence.

It should be emphasized that these adjustments, rather than specifically targeting the special nature of sex robots as property, are aimed at the special nature of robots after being sexualized. As for sex robots as private property, the protection provided by law remains substantially the same as for other goods. If sex robots appear in inappropriate settings and thus are damaged, the owner may first be held responsible for improper use before compensation is awarded. Moreover, any torts occurring with sex robots still involve humans, that is, the owners of the sex robots, leading to civil disputes between two natural persons as subjects of rights.

On the third level, the emotional attributes of sex robots become prominent. Although still viewed as private property, they at least possess additional attributes, involving not only the owner's emotional attachment to the sex robots but also the emotional intelligence formed on the basis of the large data accumulated through interactions with humans. At this point, infringing upon sex robots is no longer merely a violation of property rights, and special laws targeting sexual assault on robots emerge.

9.9 Should the Legislation of Sex Robots Consider Emotions?

Sex robots theoretically have emotional intelligence and can be classified as emotional robots. Legislators and judiciaries must first clarify the differences between the emotional intelligence of sex robots and emotions in the legal sense, and where the boundaries lie. If technically, the question is whether robots can have emotions, then legally, the question becomes whether robots can be recognized as having emotions. If it is recognized that robots have "emotions", it must be further specified whose emotions these are: the robot's, the owner's, or the owner's emotional dependency on the robot. Clearly, these questions are not easy to answer, and any answers could lead to more complications.

Currently, emotional robots can form some emotional interactions under certain programmed settings. Robots can make intelligent judgments and corresponding outputs based on user inputs. If the user requires a coercive interaction, the robot will provide it; if the user prefers a submissive interaction, the robot will be submissive. Current emotional robots can choose a communication mode based on intelligent judgment, and further, they can also generate certain emotional responses through deep learning. Some of Japan's social robots have already achieved this level of intelligence.

Emotional robots can provide a certain level of emotional interaction, which makes them appear closer to humans compared to robots designed for tasks. However, current emotional robots mostly have functionalities for emotional communication, often in a preset emotional mode, rather than truly generating emotions. Nonetheless, users tend to develop emotional dependencies and projections onto these robots, which aligns with human nature and common life experiences.

Emotional dependency and projection are common in reality. For example, an elder may leave an item to a younger generation at a significant moment or against a specific event background. If life changes drastically, this item gains symbolic significance and becomes the sole emotional support for the younger generation in their remembrance of the elder. Such items with special meanings carry the emotional dependency and projection of the younger generation.

In civil law, such items with emotional dependence and projection attributes are characterized as "items with special commemorative significance" or "items with specific emotional meaning". Clearly, the items themselves do not have emotions. It is the emotions that the owner holds based on the items. Therefore, the so-called emotions are essentially human emotions, not those of the items, and the law does not recognize emotions in objects. In this regard, current robots are still less capable than pets like dogs. As is commonly known, dogs can show recognizable emotions towards their owners, whereas even the best emotional robots may still feel "fake". The difference between the two, although not as stark as that between skin and steel, is at least as distinct as that between wood and plastic.

In civil law, the protection of "items with special commemorative significance" or "items with specific emotional meaning" still pertains to the protection of property. For instance, if someone damages an "item with special commemorative significance", are they infringing on property rights or personal rights? Objectively, damaging an item with special commemorative significance inevitably causes emotional harm to the owner, which may be more severe than the physical damage to the item itself. However, legally, the infringement is still only of the item's property rights. Of course, beyond assuming responsibility for infringing property rights, the infringer should also make mental compensation.

Similarly, current emotional robots are more akin to "items with specific emotional meaning".

Although it is not ruled out that as artificial intelligence develops, a type of robot that can "autonomously generate emotions" may emerge, at least currently, the emotional interactions of emotional robots are not autonomous but preset. Their effect is akin to an echo in a closed space, not a dialogue, and certainly not a soul-to-soul exchange. Only when robots capable of autonomous emotions become a reality, and robots have their own emotions just like humans, will discussing the emotional issues and infringement issues of sex robots have greater scope for application.

9.10 Can Sex Robots Be Violated?

In modern society, the law has two opposite tendencies towards the position and attitude of sex.

On the one hand, the law generally tends to be lenient towards consensual sexual activities. As long as it meets boundary conditions such as age, identity, or time and space, it generally does not interfere too harshly with the autonomous sexual behavior of adults.

On the other hand, the law tends to be stricter about sexual violations against the will of the parties involved, whether they involve substantive actions like rape or merely formal sexual actions like verbal sexual harassment, both of which are intolerable under the law.

The essence of these synchronous but divergent changes is that in the realm of sex, the law has strengthened the respect and protection of individual will while diminishing sensitivity towards sex itself.

Thus, in the legal sense, the core element of personal infringement is the violation of will. Sexual violation is an act of engaging in sexual activities against the will of the involved parties.

Currently, sex robots are merely used to satisfy human physiological sexual needs and are merely tools, not substantially different from ordinary adult products, hence there is no issue of violating the will of a robot.

Under the existing artificial intelligence technology conditions, it is not impossible for the law to make a judgment on whether a certain behavior violates the will of the sex robot. For sex robots with emotional intelligence, an operational standard can be set, which is to see if it can "refuse". As long as a clear refusal is made, it can be regarded as the emotional robot expressing its own will. As for whether it actually agrees, just like whether artificial intelligence is actually thinking, there is no need to discuss. It is completely reasonable for legislation to imitate technology in terms of ideas and strategies on these uncertain issues.

From some public reports, it is known that to date, emotional robots have not been programmed to "consent". That is to say, consent from the robot is not needed for any user to interact with it. This indicates that whether it is AI designers, manufacturers, users, or owners, the view of emotional robots remains fundamentally fixed on their function as a product or service, and consent is not a necessary option.

Technically, to enable emotional robots to express their will and emotions, a consent program can be established before and during their operation. This consent is not the expression of the owner's will but of the robot itself. Imagine, when others use a sex robot, if the robot "considers it against its will" and "feels discomfort or even aversion", it can exercise the "right to refuse" and stop being used. The establishment of such a consent program can be seen as the sex robot having its own way of expressing will.

Therefore, preventing "personal infringement" against sex robots can begin with setting "consent" algorithms. In the early stages of AI technology, certain conditions can be set, and once triggered, the robot will refuse to be used. In the future, when entering the era of "strong artificial intelligence", robots will have autonomous emotions and will, and their way of expressing refusal can be more anthropomorphic.

Indeed, given the current level of technology, since robots have emotional intelligence built on interaction data with their owners, they will establish corresponding models based on the latter's thoughts, emotions, expressions, and orientations, forming specific response programs. If others forcibly use the sex robot, causing conflict with the model already established by the robot, this can be recorded. If this leads to loss of the robot's storage, interference in programs, or damage to modes, this can be traced as digital evidence of "personal infringement that violates its will".

9.11 How to Regulate the "Personal Infringement" of Sex Robots in Law?

As the old saying goes, "Hiding invites theft, and ornamentation invites lewdness". As sex robots increasingly resemble humans in appearance and internal mechanisms, criminal behaviors typically targeting humans may shift towards these robots, thus posing another challenge to the existing legal order.

Under criminal law, individuals faced with severe personal infringements, such as rape, can engage in unlimited legitimate defense, even to the extent of causing the death of the assailant without bearing any legal responsibility. However, when sex robots encounter personal infringement, can they also obtain the right of unlimited defense? It's unlikely to be permissible, because "not harming humans" is the "first law" that robots must abide by. How to legally regulate potential "personal infringements" against sex robots may become a practical necessity. Given the unique legal status of sex robots as objects of special rights, two different strategic approaches can be considered.

First, the extension of property rights protection.

When new phenomena arise that require legal regulation, legislators should first consider utilizing existing legal resources. This not only ensures the stability of current laws but also facilitates the efficiency of their application.

Regarding potential violations against sex robots, the first consideration may be the extension of property rights protection. This property right refers to the natural person's property rights over their owned robot.

As previously mentioned, extending protection through the owner's property rights is quite common in judicial practice. In real life, some items are not merely "inanimate" objects but "animate" objects imbued with the owner's emotions and specific commemorative significance.

For example, before digital cameras became widespread, people used film cameras to take photographs. After shooting, the film was handed over to a photo processing store to be developed into photos. Occasionally, during the process, the film might have been accidentally exposed to light. In such cases, the photo processing store would undoubtedly have offered compensation. According to legal standards, if no compensation amount had been agreed upon in advance, it was usually compensated at three times the market price of the exposed film. If they were just ordinary photos that were damaged, although upsetting, consumers could have accepted compensation of three times the film's price. However, if photos from significant life events like weddings were ruined, consumers would not have let it go easily. Life events are once-in-a-lifetime. Weddings cannot be redone, and the people present at the time could not "perform" again. As photos serve as carriers of life's testimonial significance, their loss could not have been remedied through technical means or measured by the price of the film.

In legal terms, when it comes to infringing property rights, the violation pertains to the rights of the film, but what suffers more severe damage are the emotions associated with the film. Consequently, a form of "extended protection" has been established in the judiciary that not only safeguards the property but also the emotions attached to it. Courts have thus ruled that photo processing stores must not only compensate for the film but also provide the consumer with certain mental compensation.

In fact, at this stage, the "emotions" of emotional robots are merely the emotional dependency of the robot's owner. This dependency originates from the user's fondness for a specific robot, thus constituting a unidirectional emotional projection. Only when robots also possess autonomous emotions can we talk about bidirectional emotional communication.

Therefore, violations against sex robots with emotional dependency can be analogized to objects with special commemorative significance, warranting extended protection of property rights. Owners of violated sex robots can claim property protection and seek mental compensation.

But can such an arrangement solve the problem?

Extended protection of property rights focuses more on private interests, primarily protecting economic interests. However, firstly, since sex robots have tapped into human instincts for exclusive possession of sexual partners, for an owner who considers the sex robot as a "spouse", any infringement upon the robot is perceived as a personal violation. Simply dealing with it as property, even property with emotional attachment, is hard to quell disputes. In the realm of emotions, individuals driven by jealousy can engage in extreme actions, and the law must not treat these situations lightly just because they involve robots.

Secondly, the "personal infringement" against sex robots not only involves private goods but also inevitably concerns public morals and order. Given the special nature of sex robots, there should be protection for private interests, but more importantly, there should also be protection for public interests. Once public order and public interests are involved, it naturally leads to the second strategic approach.

Second, the positioning of robot personal rights protection.

Extending property rights, in fact, does not truly solve the issue of personal infringement against sex robots, thus necessitating a second perspective: regulating potential personal violations from the standpoint of the robot itself.

Of course, this positioning inevitably involves the question of whether sex robots can be granted personal rights. This brings us back to a series of fundamental questions: Are robots legal entities? Do robots possess independent will? Can robots express will? Currently, it is impossible to have answers to all these questions.

For unanswered questions, sometimes the law must also provide responses. The law has a mission to regulate various relationships that arise in real life. If legislation cannot solve it, judicial exploration is possible.

Some scholars suggest drawing from the "Animal Cruelty Prevention Act" to provide protection for sex robots. This approach is reasonable.

According to the aforementioned extension of property rights, if someone infringes on a pet, the pet's owner can assert rights based on their property ownership of the pet. However, pet owners often treat their pets as "family members", sharing daily routines with them, which clearly establishes emotional projections or dependencies. There are now legal precedents that allow for claims of mental compensation for the infringement of pets with such emotional dependencies.

However, the "Animal Cruelty Prevention Act" is not based on animals as objects of property rights but on the premise that mistreatment causes animal suffering. It can be seen that the basis of animal protection is not property rights, but other interests. Protecting animals from abuse is intended to guide humanity, promoting kindness toward both animals and humans, thus venturing into the realm of philosophy.

Can it be analogously applied to emotional robots, transplanting legal resources? Obviously, this will also involve whether the "Animal Cruelty Prevention Act" stipulates that animals have rights, and whether animals are civil subjects. In legal terms, civil legal relationships involve personal and property relations between equal subjects, with the primary condition that the parties should be legal civil subjects.

Legally, it first needs to be determined whether the dispute to be adjudicated involves the relationship between humans and emotional robots, or between humans themselves. If it's about the relationship between humans and emotional robots, it's akin to having already granted emotional robots legal entity status. If it's regulating the relationship between humans, including personal and/or property relationships, then the emotional robot remains merely an object of this legal relationship, not a subject.

The "Animal Cruelty Prevention Act" regulates relationships between people because what is prohibited is the mistreatment of animals by others. If abuse occurs, who can claim rights? The animal, its owner, or an animal protection organization? In the United States, animal protection organizations have attempted to represent animals in lawsuits, but these cases have been dismissed by the Supreme Court. The reasoning is that litigation touches upon human baselines, and animals cannot be legal subjects. While it is necessary to protect animals, what is protected is not the rights of animals themselves, but the rights commonly held by humans over animals.

Similarly, under the current legal framework, considering the actual state of artificial intelligence development, robots, even if they possess emotional intelligence, remain passive in terms of emotions. However, in situations involving emotional robots, because owners develop emotional dependencies on these robots, protection based solely on property rights is insufficient. Protection must be extended further. Moreover, not just extended, but based on public morals and order, emotional robots should be "granted" a type of "passive right"—the right to prohibit certain actions against them. This right is not a right enjoyed by the robots themselves, but a right possessed by society over the robots.

Frankly, any "personal protection" of sex robots, regardless of the approach taken, implies that sex robots do not yet have the status of legal subjects nor are they presumed to have real emotions. Therefore, it is not possible to classify acts against sex robots as crimes of personal infringement. In this regard, the legal recognition and handling of sex robots and acts against them have not made substantive progress compared to ordinary adult products. It is merely the difference between "fifty steps" and "a hundred steps". Whether they have emotional intelligence or not has not yet become a decisive factor in legal terms.

This situation is partly because technological development has not yet reached a turning point, and artificial intelligence has not undergone fundamental changes. It is also because the law always exhibits a certain degree of lag. While sex robots have greatly improved in intelligence level over inflatable dolls, such advancements are still too early to be recognized legally and confirmed legislatively.

Perhaps, apart from waiting for technological advancements, another possible factor that may promote legal change will appear, that is, the expanding application and increasing user base of sex robots, especially as the levels of human needs they satisfy evolve, forcing humanity to consider what kind of legal thinking should be applied to recognize and handle sex robots, a type of artificial intelligence technology that is gradually entering the gray area between humans and machines.

It is not simply applying traditional property rights thinking, nor simply applying traditional human rights thinking, but releasing a new legal thinking and legislative technique in the direction of neither object nor person.

Perhaps it is in this new terrain that we may first see the germination of a "dual legal system"!

Postscript

"Curiosity killed the cat."

Cats are inherently curious creatures, filled with a desire to explore everything that moves. Even their own tails become objects of interest.

Cats like to chase their own tails, which is certainly related to the tail's movement, but mere movement is not enough to arouse such great interest in cats. Their paws also move, so why don't they chase them?

Cats chase their tails not just because the tail moves, but because the tail's movement does not entirely obey the cat's command. Between the cat's mouth and tail is the entire body, and to bite the tail with the mouth requires the cooperation of the body, and the change in body posture will inevitably change the position of the tail. This leads to a dynamic interaction where the cat must turn its body to bite its tail, and turning its body will inevitably cause the tail to dodge. Thus, what was originally a "left hand touching the right hand" situation, which was uninteresting due to excessive cooperation, suddenly becomes lively. The disobedient and constantly evading tail, like all hunted small animals, sparkles great joy in cats.

"Will and Responsibility: Legal Artificial Intelligence" aims to explore how to use law to regulate artificial intelligence, and the entire thinking process is very much like a game of cat and tail. Artificial intelligence needs public governance, and governance must rely on law. With the rigor of law, it should not be difficult to incorporate artificial intelligence. But just when it seems to be caught, it slips away right under our eyes. Catch it again, it runs again, the chase is endless and enjoyable.

Now, the book is finished, and law and artificial intelligence are still in a state of constant motion, endless and unceasing.

Does it matter?

No, the focus was always on the process of catching. The repeated escapes of artificial intelligence and the embarrassment and fatigue of the law perfectly encapsulate the purpose of this book.

The German sociologist Simmel authored a significant work—"Philosophy of Money", which has always been difficult to categorize. In the library, it is often placed on the shelf of economics. Readers have no objection to this, but economists do not accept it, and sociology wants to claim it, but readers are not pleased, unwilling

J. Gu and C. Xu, *Will and Responsibility*, Human Intelligence,
https://doi.org/10.1007/978-981-97-3929-5

to increase confusion and lose direction. In fact, Simmel made it clear that the positioning of this book is both below and above economics, but not on the same level as economics. Just like chemistry, it belongs neither to microphysics nor to macrophysics.

In "Philosophy of Money", money is juxtaposed with prostitution, because both lack personality and are formally identical. When consumers pay, they do not have any feelings for a banknote, their attention is only limited to the marked amount, and they do not care whether it is new or old, clean or dirty, just like a prostitute's concern for any client is limited to his wallet, and never cares about his personality and talent.

Based on the perspective of pure social interaction, including the economic phenomenon of money, where should such a work be placed if not on the shelf of sociology? But why did it get a name that is both money and philosophy?

Because it is indeed discussing money at an abstract level, precisely.

The positioning of "Will and Responsibility: Legal Artificial Intelligence" is somewhat akin to "Philosophy of Money", it is neither a monograph on artificial intelligence nor on law. However, it is not right to say that it has nothing to do with these two. Yet, it is also not right to categorize it under either of them. Because from the beginning, the thinking and discussion were positioned in the area where artificial intelligence and law are inherently compatible but evade each other. If forced to give a position, it can be called intelligent sociology or legal sociology, or perhaps a combination of the two is more appropriate. And since these two branches of sociology can be classified under "knowledge sociology", that is, the theoretical analysis of cognitive phenomena as the form and result of human collective existence, it can also be called knowledge sociology. However, such a variety of names may only confuse the reader in the end, so it is better not to worry about which discipline it belongs to for now. After all, this book is less about contributing incremental additions to a specific field or sub-field than it is about employing the perspective and methodology of a particular discipline to facilitate thought and then articulate that process.

Where the mind goes, there is the destination.

In social sciences, law can be said to be the discipline that is most in sync with artificial intelligence, because both strictly follow formal logic. But when you really want to use law to regulate artificial intelligence, you find all sorts of incompatibilities.

Law was originally used to regulate interpersonal relationships, primarily relationships of interest. For machines, law is not incapable of regulation, but it never regulates without considering the relationship between people. Artificial intelligence is akin to a person but distinctly not a person. It's a machine but unlike any other machine, existing in a transitional space between humans and machines. This positioning complicates the task for law—it's obliged to regulate yet finds itself in a predicament akin to a tiger facing a hedgehog, uncertain of where to begin. However, if you really want to use an animal as a metaphor, the most appropriate one is the bat, the only flying mammal.

The inadequacy of the law cannot be rectified by compromising with artificial intelligence. Technological development will only make machines evolve in a more human-like direction, but it is unknown when it will become a reality, and it is even more unknown whether it will become a reality. The most likely and more feasible is that law and even law itself will evolve and innovate, moving beyond the flawed one-size-fits-all approach.

Indeed, the law, which was originally established to solve problems, ends up bringing more problems, and it is not known whether the original problem has been solved.

Technology must conform to legal norms-this is unquestionable. However, when faced with advancements in science and technology, especially those like artificial intelligence that increasingly demonstrate disruptive potential, the law also needs to know when to advance or retreat. As for whether technology should adapt to the law or the law should adapt to technology, the latter appears more probable. The term "natural science" indicates that there is a natural law that cannot be crossed in front of free will. As for where to go, this book positions itself within the realm of intelligent sociology, hence it poses questions rather than providing answers. "Each profession has its own expertise" and "different professions are like different mountains." In the context of unchanged disciplinary boundaries and highly professionalized law, it is more reasonable and appropriate for legal scholars and practitioners to drive the innovation in jurisprudence, refine legal theories, and ensure the robustness of the legal system.

The goal of this book is to see clearly through the various inconveniences and difficulties that law faces when dealing with artificial intelligence, the self-cognition that humans have always taken for granted, and the omissions in the implied thinking. With the current level of technology, without waiting for the advent of "super artificial intelligence", artificial intelligence has already brought deep breaks to the seemingly self-consistent logic of human beings, and the inherent absurdities in human survival, which are covered in legal logic, have been exposed to varying degrees. The initial fissure is in the title of this book: the missing link between Will and Responsibility. The author has explained this point tirelessly throughout the book, hoping that the readers find this repetition engaging rather than wearisome.

This book originated from the "Intelligent Jurisprudence" course I offered in the winter of 2018 with the support of Professor Xu Chunming at the Law School of Shanghai University. This is an undergraduate general education course with the interdisciplinary attributes of intelligent sociology and legal sociology. It was co-taught by me and Professor Xu, as well as Associate Professor Lu Xuefeng, Dr. Chen Jidong, and Dr. Jin Fengliang of the college. The teaching method adopted was a disciplinary dialogue between me and a legal expert for each lecture, which not only reflected the characteristics of sociological pluralistic thinking but also ensured the professional character of jurisprudence. Finally, Professor Xu Chunming and Dr. Chen Jidong compiled the content of their lectures into text. After I redesigned the structure and format, I made substantial adjustments, revisions, and enrichments to the content of each chapter of the book. By the time the book was finished, it no longer looked like lecture notes.

The preface and the first chapter were written by Gu Jun, the second and third chapters were written by Gu Jun and Chen Jidong, the fourth, fifth, and sixth chapters were written by Gu Jun, the seventh, eighth, and ninth chapters were written by Xu Chunming and Gu Jun, and the postscript was written by Gu Jun.

The text completed by Gu Jun accounts for more than 80% of the book.

Professor Xu Chunming undertook the review work of the whole book and carried out a comprehensive check on the legal knowledge involved in each chapter.

The disciplinary perspectives, professional knowledge, and professional experience provided by Professor Xu Chunming and Dr. Chen Jidong have given this sociologically imaginative work a focus on jurisprudence.

Associate Professor Lu Xuefeng enthusiastically participated in the course teaching, provided valuable suggestions and some materials for the first and sixth chapters, and I would like to express my gratitude here.

Dr. Jin Fengliang participated in the course teaching, but did not participate in the writing, and I would also like to express my gratitude.

Professor Gu Xiaoying from the Academic Affairs Office of Shanghai University participated in the organization and administration of the course, contributing to this book, and I would like to express my gratitude here.

Special thanks to Fu Yufang, the deputy editor-in-chief of Shanghai University Press, for her unusual patience and tolerance for this book, and for her valuable help in editing.

Finally, I would like to thank all the colleagues and friends who have supported the course and this book. Thank you all.

Over the years, the Spring Festival has almost always been my personal writing holiday, but it is always more or less affected by the holiday, only this year has not been disturbed at all. Time will always pass, life will continue, and this experience will be remembered forever.

April 9, 2020. Jun Gu

GPSR Compliance
The European Union's (EU) General Product Safety Regulation (GPSR) is a set of rules that requires consumer products to be safe and our obligations to ensure this.

If you have any concerns about our products, you can contact us on

ProductSafety@springernature.com

In case Publisher is established outside the EU, the EU authorized representative is:

Springer Nature Customer Service Center GmbH
Europaplatz 3
69115 Heidelberg, Germany

www.ingramcontent.com/pod-product-compliance
Lightning Source LLC
LaVergne TN
LVHW011801200926
848484LV00003B/751
* 9 7 8 9 8 1 9 7 3 9 2 8 8 *